Gasification of Unconventional Feedstocks

Gasification of Unconventional Feedstocks

James G. Speight PhD, DSc
CD&W Inc.,
Laramie, Wyoming, USA

AMSTERDAM • BOSTON • HEIDELBERG • LONDON
NEW YORK • OXFORD • PARIS • SAN DIEGO
SAN FRANCISCO • SINGAPORE • SYDNEY • TOKYO

Gulf Professional Publishing is an imprint of Elsevier

ELSEVIER

Gulf Professional Publishing is an imprint of Elsevier
25 Wyman Street, Waltham, MA 02451, USA
The Boulevard, Langford Lane, Kidlington, Oxford, OX5 1GB, UK

Notices
Knowledge and best practice in this field are constantly changing. As new research
and experience broaden our understanding, changes in research methods or professional
practices, may become necessary.

Practitioners and researchers must always rely on their own experience and knowledge in
evaluating and using any information or methods described herein. In using such information
or methods they should be mindful of their own safety and the safety of others, including
parties for whom they have a professional responsibility.

To the fullest extent of the law, neither the Publisher nor the authors, contributors, or editors,
assume any liability for any injury and/or damage to persons or property as a matter of
products liability, negligence or otherwise, or from any use or operation of any methods,
products, instructions, or ideas contained in the material herein.

Library of Congress Cataloging-in-Publication Data
A catalog record for this book is available from the Library of Congress

British Library Cataloguing-in-Publication Data
A catalogue record for this book is available from the British Library.

ISBN: 978-0-12-799911-1

For information on all Gulf Professional Publishing publications
visit our website at store.elsevier.com

This book has been manufactured using Print On Demand technology. Each copy is
produced to order and is limited to black ink. The online version of this book will show
color figures where appropriate.

Working together
to grow libraries in
developing countries

ELSEVIER | Book Aid International

www.elsevier.com • www.bookaid.org

Transferred to Digital Printing in 2014

CONTENTS

PREFACE

Gasification processes can accept a variety of feedstocks but the reactor must be selected on the basis of the feedstock properties and behavior in the process, especially when coal, biomass, and various wastes are considered as gasification feedstocks.

The projections for the continued use of fossil fuels indicate that there will be at least another five decades of fossil fuel use (especially coal and petroleum) before biomass and other forms of alternative energy take a firm hold, although significant inroads are being made into the gasification of various feedstocks. However, the ever-increasing global energy demand and the fast depleting fossil fuels have shifted focus on sustainable energies such as biomass and waste in the recent past. The importance of the gasification of such alternative feedstocks cannot be under-appreciated as potential sources of sustainable energy to meet the energy demands of future generations. The various technologies that are currently in practice at the commercial and pilot scale, with respect to bubbling, circulating fluidized beds and dual fluidized beds, are being developed for feedstocks other than coal.

In preparation for the depletion of fossil fuel resources, gasification can be proposed as a viable and reliable alternative solution for energy recovery from a variety of feedstocks. However, the process still faces some technical and economic problems, mainly related to the highly heterogeneous nature of unconventional feedstocks such as biomass and municipal solid wastes and the relatively limited number of gasification plants worldwide based on this technology that have continuous operating experience under commercial conditions.

It is the purpose of this book to summarize major issues related to gasification technology and to relate how the continually evolving technology will play a role in future energy production.

Dr. James G. Speight
Laramie, Wyoming, USA
October 2013

CHAPTER 1

Feedstocks

1 INTRODUCTION

Gasification is a process that converts organic carbonaceous feedstocks into carbon monoxide, carbon dioxide, and hydrogen by reacting the feedstock at high temperatures ($>700°C$, $1290°F$), without combustion, with a controlled amount of oxygen and/or steam (Lee et al., 2007; Speight, 2008, 2013a, 2013b). The resulting gas mixture (*synthesis gas*) is called a producer gas and is itself a fuel. The power derived from carbonaceous feedstocks and gasification followed by the combustion of the product gas(es) is considered to be a source of renewable energy if the gaseous products are from a source (e.g., biomass) other than a fossil fuel (Speight, 2008).

For many decades coal has been the primary feedstock for gasification units but recent concerns about the use of fossil fuels and the resulting environmental pollutants, irrespective of the various gas cleaning processes and gasification plant environmental cleanup efforts, there is a move to feedstocks other than coal for gasification processes (Speight, 2013a, 2013b). But more pertinent to the present text, the gasification process can also use carbonaceous feedstocks which would otherwise have been discarded and unused, such as waste biomass and other similar biodegradable wastes. Various feedstocks such as biomass, petroleum resids, and other carbonaceous wastes can be used to their fullest potential. In fact, the refining industry has seen fit to use residua gasification as a source of hydrogen for the past several decades (Speight, 2014).

The advantage of the gasification process when a carbonaceous feedstock (a feedstock containing carbon) or hydrocarbonaceous feedstock (a feedstock containing carbon and hydrogen) is employed is that the product of focus − synthesis gas − is potentially more useful as an energy source and results in an overall cleaner process. The production of synthesis gas is a more efficient production of an energy source than, say, the direct combustion of the original feedstock because synthesis gas can be (1) combusted at higher temperatures,

(2) used in fuel cells, (3) used to produce methanol, (4) used as a source of hydrogen, and particularly (5) converted via the Fischer − Tropsch process into a range of synthesis liquid fuels suitable for use in gasoline engines or diesel engines (Chapters 4, and 5).

Gasification plants are cleaner because fewer sulfur and nitrogen byproducts are produced thereby contributing to a *decrease* in smog formation and acid rain deposition. For this reason, gasification is an appealing process for the utilization of relatively inexpensive feedstocks that might otherwise be declared as waste and sent to a landfill (where the production of methane − a so-called greenhouse gas − will be produced) or combusted which may not (depending upon the feedstock) be energy efficient. Overall, use of a gasification technology (Chapter 3) with the necessary gas cleanup options can have a smaller environmental footprint and lesser effect on the environment than landfill operations or combustion of the waste. Indeed, the increasing interest in gasification technology reflects a convergence of two changes in the electricity generation marketplace: (1) the maturity of gasification technology, and (2) the extremely low emissions from integrated gasification combined cycle (IGCC) plants, especially air emissions, and the potential for lower cost control of greenhouse gases than other coal-based systems (Speight, 2013b).

Liquid fuels, including gasoline, diesel, naphtha and jet fuel, are usually processed by refining crude oil (Speight, 2014). Due to the direct distillation, crude oil is the raw material most suited to liquid fuel production. However, with the fluctuating and rising cost of processing petroleum, coal-to-liquids (CTL) and biomass-to-liquids (BTL) processes are employed as alternative routes for liquid fuel production. Both of these feedstocks are converted to synthesis gas − a mixture of carbon monoxide and hydrogen − after which the tried-and-tested Fischer − Tropsch (FT) technology is used to convert the synthesis gas to a mixture of liquid products, which is further upgraded using known petroleum refinery technologies to produce gasoline, naphtha, diesel fuel, and jet fuel (Dry, 1976; Mukoma et al., 2006; Chadeesingh, 2011; Speight, 2014).

This chapter presents a description of the various non-coal feedstocks that can be used in gasification to produce a variety of gases, including synthesis gas (a mixture of carbon monoxide and hydrogen).

2 FEEDSTOCKS

Gasification processes can accept a variety of feedstocks but the reactor must be selected on the basis of the feedstock properties and behavior in the process, especially when coal and biomass are considered as gasification feedstocks (Table 1.1). Furthermore, because of the historical use of coal for gasification purposes (Speight, 2013a), it is the feedstock against which the suitability of all other feedstocks is measured. Therefore, inclusion of coal among the gasification feedstocks in this section is warranted.

2.1 Coal

Coal is a combustible organic sedimentary rock (composed primarily of carbon, hydrogen, and oxygen) formed from ancient vegetation and consolidated between other rock strata to form coal seams. The harder forms, such as anthracite coal, can be regarded as organic metamorphic rocks because of a higher degree of maturation (Speight, 2013a).

Coal is the largest single source of fuel for the generation of electricity worldwide (EIA, 2007; Speight, 2013b), as well as the largest source of carbon dioxide emissions, which have been implicated (rightly or wrongly) as the primary cause of global climate change (Speight, 2013b). Many of the proponents of global climate change forget (or

Table 1.1 Comparison of the Properties of Coal and Biomass

Property	Bio mass	Coal
Fuel density (kg/m^3)	~500	~1300
Particle size	~3 mm	~100 μm
C content (wt% of dry fuel)	42−54	65−85
O content (wt% of dry fuel)	35−45	2−15
S content (wt% of dry fuel)	max 0.5	0.5−7.5
SiO$_2$ content (wt% of dry ash)	23−49	40−60
K$_2$O content (wt% of dry ash)	4−48	2−6
Al$_2$O$_3$ content (wt% of dry ash)	2.4−9.5	15−25
	1.5−8.5	8−18
Fe$_2$O$_3$ content (wt% of dry ash)	418−426	490−595
Ignition temperature (K)	560−575	−
Peak temperature (K)	Low	high
Friability	14−21	23−28

From (Demirbaş 2005).

refuse to acknowledge) that the earth is in an interglacial period when warming and climate change can be expected – this was reflected in the commencement of the melting of the glaciers approximately 11 000 years ago. Thus, taking into account the geological sequence of events, the contribution of carbon dioxide from anthropogenic sources is not known with any degree of accuracy.

Coal occurs in different forms or *types* (Speight, 2013a). Variations in the nature of the source material and local or regional variations in the coalification processes caused the vegetal matter to evolve differently. Thus, various classification systems exist to define the different types of coal. Thus, as geological processes increase their effect over time, the coal precursors are transformed over time into:

(1) Lignite – also referred to as brown coal and is the lowest rank of coal that is used almost exclusively as fuel for steam-electric power generation; *jet* is a compact form of lignite that is sometimes polished and has been used as an ornamental stone since the Iron Age.
(2) Sub-bituminous coal, which exhibits properties ranging from those of lignite to those of bituminous coal and are used primarily as fuel for steam-electric power generation.
(3) Bituminous coal, which is a dense coal, usually black, sometimes dark brown, often with well-defined bands of bright and dull material, used primarily as fuel in steam-electric power generation, with substantial quantities also used for heat and power applications in manufacturing and to make coke.
(4) Anthracite, which is the highest rank coal and is a hard, glossy, black coal used primarily for residential and commercial space heating.

Chemically, coal is a hydrogen-deficient hydrocarbon with an atomic hydrogen-to-carbon ratio near 0.8, as compared to petroleum hydrocarbons, which have an atomic hydrogen-to-carbon ratio approximately equal to 2, and methane (CH_4) that has an atomic carbon-to-hydrogen ratio equal to 4. For this reason, any process used to convert coal to alternative fuels must add hydrogen or redistribute the hydrogen in the original coal to produce hydrogen-rich products and coke (Speight, 2013a).

In the gasification of coal, a mixture of gases is produced – in addition to synthesis gas (mixtures of carbon monoxide and hydrogen), methane and other hydrocarbons are also produced depending on the conditions involved. The gases are withdrawn and may be burned to produce heat, generate electricity, or are used as synthesis gas in indirect liquefaction or the production of chemicals. The indirect liquefaction process involves Fischer − Tropsch synthesis which results in the production of hydrocarbons that can be refined into liquid fuels.

By developing adsorbents to capture the pollutants (mercury, sulfur, arsenic, and other harmful gases), there is a focus not only on the reduction of the quantity of gaseous emissions but also on maximization of the thermal efficiency of the cleanup methods. Thus, gasification offers one of the most clean and versatile ways to convert the energy contained in coal into electricity, hydrogen, and other sources of power.

However, coal is an abundant natural resource but the combustion or gasification of coal produces both toxic pollutants and greenhouse gases. By developing adsorbents to capture the pollutants (mercury, sulfur, arsenic, and other harmful gases), there have been serious efforts to reduce the quantity of emitted gases and also to maximize the thermal efficiency of the cleanup. Thus, gasification can offer one of the most clean and versatile ways to convert the energy contained in coal into electricity, hydrogen, and other sources of power.

2.2 Petroleum Coke

Coke is the solid carbonaceous material produced from petroleum during thermal processing. More particularly, coke is the residue left by the destructive distillation (i.e., thermal cracking such as the delayed coking process) of petroleum residua. The coke formed in catalytic cracking operations is usually non-recoverable because of the materials deposited on the catalyst during the process and such coke is often employed as fuel for the process (Speight, 2014). It is often characterized as a solid material with a honeycomb-type of appearance having high carbon content (95% + w/w) with some hydrogen and, depending on the process, sulfur and nitrogen as well. The color varies from gray to black, and the material is insoluble in organic solvents.

Typically, the composition of petroleum coke varies with the source of the crude oil, but in general, large amounts of high-molecular-weight complex hydrocarbons (rich in carbon but correspondingly poor in hydrogen) make up a high proportion. The solubility of petroleum *coke* in carbon disulfide has been reported to be as high as 50 to 80%, but this is in fact a misnomer and is due to soluble product adsorbed on the coke – by definition coke is the insoluble, honeycomb material that is the end product of thermal processes. However, coke is not always a product with little use – three physical structures of coke can be produced by delayed coking: (1) shot coke, (2) sponge coke, or (3) needle coke, which find different uses within the industry.

Shot coke is an abnormal type of coke resembling small balls. Due to mechanisms not well understood the coke from some coker feedstocks forms into small, tight, non-attached clusters that look like pellets, marbles, or ball bearings. It usually is a very hard coke, i.e., low Hardgrove grindability index (Speight, 2013a). Such coke is less desirable to the end users because of difficulties in handling and grinding. It is believed that feedstocks high in asphaltene constituents and low API favor shot coke formation. Blending aromatic materials with the feedstock and/or increasing the recycle ratio reduces the yield of shot coke. Fluidization in the coke drums may cause formation of shot coke. Occasionally, the smaller *shot coke* may agglomerate into ostrich egg sized pieces. Such coke may be more suitable as a gasification feedstock.

Sponge coke is the common type of coke produced by delayed coking units. It is in a form that resembles a sponge and has been called honeycombed. Sponge coke, mostly used for anode-grade carbon, is dull and black, having porous, amorphous structure.

Needle coke (*acicular coke*), a special quality coke produced from aromatic feed stocks, is silver-gray, having crystalline broken needle structure, and is believed to be chemically produced through cross-linking of condensed aromatic hydrocarbons during coking reactions. It has a crystalline structure with more unidirectional pores and is used in the production of electrodes for the steel and aluminum industries and is particularly valuable because the electrodes must be replaced regularly.

Petroleum coke is employed for a number of purposes, but its chief use is (depending upon the degree of purity, i.e., contains a low amount of contaminants) for the manufacture of carbon electrodes for aluminum refining, which requires a high-purity carbon – low in ash and sulfur free; the volatile matter must be removed by calcining.

In addition to its use as a metallurgical reducing agent, petroleum coke is employed in the manufacture of carbon brushes, silicon carbide abrasives, and structural carbon (e.g., pipes and Raschig rings), as well as calcium carbide manufacture from which acetylene is produced:

$$\text{Coke} \rightarrow \text{CaC}_2$$
$$\text{CaC}_2 + \text{H}_2\text{O} \rightarrow \text{HC} \equiv \text{CH}$$

(1.1.1)

Considering the properties of coke and the potential non-use of the highly contaminated material, gasification is the only technology which makes possible for the refineries the zero residue target, contrary to all conversion technologies, thermal cracking, catalytic cracking, cooking, deasphalting, hydroprocessing, etc., which can only reduce the bottom volume, with the complication that the residue qualities generally get worse with the degree of conversion (Speight, 2014).

Indeed, the flexibility of the gasification technology permits the refinery to handle any kind of refinery residue, including petroleum coke, tank bottoms, and refinery sludge and makes available a range of value added products, electricity, steam, hydrogen and various chemicals based on synthesis gas chemistry: methanol, ammonia, MTBE, TAME, acetic acid, and formaldehyde (Speight, 2008, 2013a). With respect to gasification, no other technology processing low value refinery residues can come close to the emission levels achievable with gasification (Speight, 2014) and is projected to be a major part of the refinery of the future (Speight, 2011b).

Gasification is also a method for converting petroleum coke and other refinery non-volatile waste streams (often referred to as *refinery residuals* and include but not limited to atmospheric residuum, vacuum residuum, visbreaker tar, and deasphalter pitch) into power, steam, and hydrogen for use in the production of cleaner transportation fuels. For the gasification of coal and biomass, the main requirement for a feedstock to a gasification unit is that it contains both hydrogen and carbon (Table 1.2).

The typical gasification system incorporated into the refinery consists of several process plants including (1) feed preparation, (2) the gasifier, (3) an air separation unit (ASU), (4) synthesis gas cleanup, (5) sulfur recovery unit (SRU), and (6) downstream process options such as Fischer − Tropsch synthesis (FTS) and methanol synthesis (MTS), depending on the desired product state (Figure 1.1).

Table 1.2 Types of Refinery Feedstocks Available for Gasification on-site					
	Units	Vacuum Residuum	Visbreaker Tar	Asphalt	Petcoke
Ultimate analysis					
C	wt/wt	84.9%	86.1%	85.1%	88.6%
H	"	10.4%	10.4%	9.1%	2.8%
N[a]	"	0.5%	0.6%	0.7%	1.1%
S[a]	"	4.2%	2.4%	5.1%	7.3%
O	"		0.5%		0.0%
Ash	"	0.0%		0.1%	0.2%
Total	wt/wt	100.0%	100.0%	100.0%	100.0%
H_2/C ratio	mol/mol	0.727	0.720	0.640	0.188
Density					
Specific gravity	$60^a/60^a$	1.028	1.008	1.070	0.863
API gravity	API^a	6.2	8.88	0.8	–
Heating values					
Higher heating value (dry)	M Btu/lb	17.72	18.6	17.28	14.85
Lower heating value (dry)	"	16.77	17.6	16.45	14.48

[a]*Nitrogen and Sulfur contents vary widely.*
Source: National Energy Technology Laboratory, United States Department of Energy, Washington, DC.
http://www.netl.doe.gov/technologies/coalpower/gasification/gasifipedia/7-advantages/7-3-4_refinery.html.

The benefits to a refinery for adding a gasification system for petroleum coke or other residuals are: (1) production of power, steam, oxygen, and nitrogen for refinery use or sale, (2) source of synthesis gas for hydrogen to be used in refinery operations and for the production of light refinery products through Fischer – Tropsch synthesis, (3) increased efficiency of power generation, improved air emissions, and reduced waste stream versus combustion of petroleum coke or residua or incineration, (4) no off-site transportation or storage for petroleum coke or residuals, and (5) the potential to dispose of waste streams including hazardous materials.

Gasification of coke can provide high-purity hydrogen for a variety of uses within the refinery such as (1) sulfur removal, (2) nitrogen removal, as well as removal of other impurities from intermediate to finished product streams and in hydrocracking operations for the conversion of heavy distillates and oils into light products, naphtha, kerosene, and diesel fuel (Speight, 2014). Hydrocracking and severe hydrotreating require hydrogen which is at least 99% v/v pure,

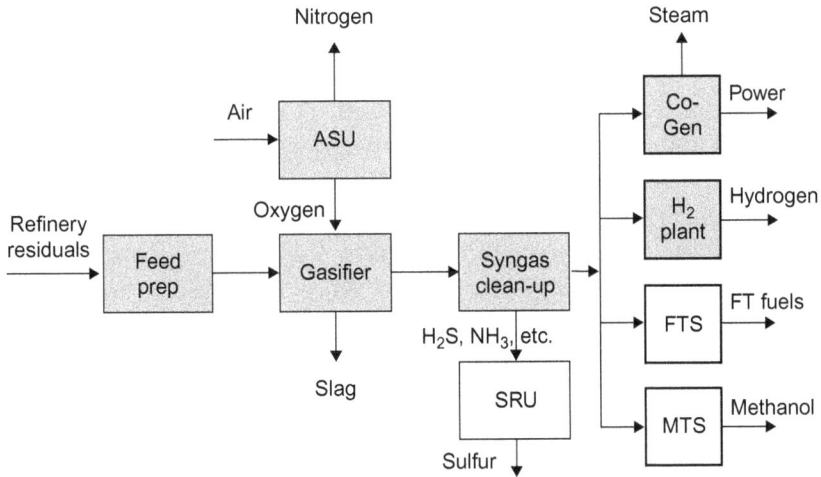

Figure 1.1 Gasification as might be employed on-site in a refinery. Source: National Energy Technology Laboratory, United States Department of Energy, Washington, DC. http://www.netl.doe.gov/technologies/ coalpower/gasification/gasifipedia/7-advantages/7-3-4_refinery.html

while less severe hydrotreating requires a gas stream containing hydrogen on the order of 90% v/v purity.

Electric power and high pressure steam can be generated by the gasification of petroleum coke and residuals to drive mostly small and intermittent loads such as compressors, blowers, and pumps. Steam can also be used for process heating, steam tracing, partial pressure reduction in fractionation systems, and stripping low-boiling components to stabilize process streams.

During gasification some soot (typically 99% + carbon) is produced, which ends up in the quench water. The soot is transferred to the feedstock by contacting, in sequence, the quench water blowdown with naphtha, and then the naphtha – soot slurry with a fraction of the feed. The soot mixed with the feed is recycled to the gasifier, thus achieving 100% conversion of carbon to gas.

2.3 Petroleum Residua

A petroleum *resid* (*residuum*, pl. *residua*, *resids*) is the residue obtained from petroleum after non-destructive distillation of the petroleum feedstock has removed all of the volatile materials. The temperature of the distillation is usually maintained below 350°C (660°F) since the rate of thermal decomposition of petroleum constituents is minimal below this temperature but the rate of thermal decomposition of petroleum constituents is substantial above 350°C (660°F) (Speight, 2014).

Resids are black, viscous materials and are obtained by distillation of a crude oil under atmospheric pressure (atmospheric residuum) or under reduced pressure (vacuum residuum). They may be liquid at room temperature (generally atmospheric residua) or almost solid (generally vacuum residua) depending upon the nature of the crude oil (Speight, 2014). When a residuum is obtained from a crude oil and thermal decomposition has commenced, it is more usual to refer to this product as *pitch* – although this term is usually applied to the non-volatile product from coal tar. The differences between the parent petroleum and the (atmospheric and vacuum) residua are due to the relative amounts of various constituents present, which are removed from or remain in the non-volatile residuum by virtue of their relative volatility.

The chemical composition of a residuum from an asphaltic crude oil is complex. Physical methods of fractionation usually indicate high proportions of asphaltene constituents (heptane-insoluble materials) and resin constituents (simply, heptane-soluble materials but propane-insoluble materials), even in amounts up to 50% (or higher) of the residuum. In addition, the presence of ash-forming metallic constituents, including such organometallic compounds as those of vanadium and nickel, is also a distinguishing feature of residua and the heavier oils. Furthermore, the deeper the *cut* into the crude oil, the greater is the concentration of sulfur and metals in the residuum and the greater the deterioration in physical properties (Speight, 2014).

2.4 Asphalt, Tar, and Pitch

Asphalt does not occur naturally but is manufactured from petroleum and is a black or brown material that has a consistency varying from a viscous liquid to a glassy solid (Speight, 2014). To a point, asphalt can resemble bitumen (isolated from tar sand formation), hence the tendency to refer to bitumen (incorrectly) as *native asphalt*. It is recommended that there be differentiation between asphalt (manufactured) and bitumen (naturally occurring) other than by use of the qualifying terms *petroleum* and *native* since the origins of the materials may be reflected in the resulting physicochemical properties of the two types of materials. It is also necessary to distinguish between the asphalt which originates from petroleum by refining and the product in which the source of the asphalt is a material other than petroleum, e.g., *Wurtzilite asphalt* (Speight, 2014). In the absence of a qualifying word, it should be assumed that the word *asphalt* (with or without qualifiers such as *cutback*, *solvent*, and *blown*, which indicate the process used to produce the asphalt) refers to the product manufactured from petroleum.

When the asphalt is produced simply by distillation of an asphaltic crude oil, the product can be referred to as *residual asphalt* or *straight-run asphalt*. For example, if the asphalt is prepared by *solvent* extraction of residua or by light hydrocarbon (propane) precipitation, or if *blown* or otherwise treated, the term should be modified accordingly to qualify the product (e.g., *solvent asphalt*, *propane asphalt*, *blown asphalt*).

Asphalt softens when heated and is elastic under certain conditions and has many uses. For example, the mechanical properties of asphalt are of particular significance when it is used as a binder or adhesive. The principal application of asphalt is in road surfacing, which may be done in a variety of ways. Light oil *dust layer* treatments may be built up by repetition to form a hard surface, or a granular aggregate may be added to an asphalt coat, or earth materials from the road surface itself may be mixed with the asphalt. Other important applications of asphalt include canal and reservoir linings, dam facings, and sea works. The asphalt so used may be a thin, sprayed membrane, covered with earth for protection against weathering and mechanical damage, or thicker surfaces, often including riprap (crushed rock). Asphalt is also used for roofs, coatings, floor tiles, soundproofing, waterproofing, and other building-construction elements and in a number of industrial products, such as batteries. For certain applications an asphaltic emulsion is prepared, in which fine globules of asphalt are suspended in water.

Tar is a product of the destructive distillation of many bituminous or other organic materials and is a brown to black, oily, viscous liquid to semisolid material. However, *tar* is most commonly produced from *bituminous coal* and is generally understood to refer to the product from coal, although it is advisable to specify *coal tar* if there is the possibility of ambiguity. The most important factor in determining the yield and character of the coal tar is the carbonizing temperature. Three general temperature ranges are recognized, and the products have acquired the designations: *low-temperature tar* (approximately 450 to 700°C; 540 to 1290°F); *mid-temperature tar* (approximately 700 to 900°C; 1290 to 1650°F); and *high-temperature tar* (approximately

900 to 1200°C; 1650 to 2190°F). Tar released during the early stages of the decomposition of the organic material is called *primary tar* since it represents a product that has been recovered without the secondary alteration that results from prolonged residence of the vapor in the heated zone.

Treatment of the distillate (boiling up to 250°C, 480°F) of the tar with caustic soda causes separation of a fraction known as *tar acids*; acid treatment of the distillate produces a variety of organic nitrogen compounds known as *tar bases*. The residue left following removal of the heavy oil, or distillate, is *pitch*, a black, hard, and highly ductile material.

In the chemical-process industries, pitch is the black or dark brown residue obtained by distilling coal tar, wood tar, fats, fatty acids, or fatty oils.

Coal tar pitch is a soft to hard and brittle substance containing chiefly aromatic resinous compounds along with aromatic and other hydrocarbons and their derivatives; it is used chiefly as road tar, in waterproofing roofs and other structures, and to make electrodes. *Wood tar pitch* is a bright, lustrous substance containing resin acids; it is used chiefly in the manufacture of plastics and insulating materials and in caulking seams. *Pitch* derived from fats, fatty acids, or fatty oils by distillation are usually soft substances containing polymers and decomposition products; they are used chiefly in varnishes and paints and in floor coverings.

Any of the above derivatives can be used as a gasification feedstock. The properties of asphalt change markedly during the aging process (oxidation in service) to the point where the asphalt fails to perform the task for which it was designed. In some cases, the asphalt is recovered and re-processed for additional use or it may be sent to a gasifier.

2.5 Tar Sand Bitumen

Tar sand bitumen is used interchangeably with the term oil sand bitumen in Canada and the word bitumen (also, on occasion, incorrectly referred to as native asphalt, and extra heavy oil) includes a wide variety of reddish brown to black materials of semisolid, viscous to brittle character that can exist in nature with no mineral impurity or with a mineral matter content that exceeds 50% w/w (Speight, 2009, 2013c,

2014). Bitumen is frequently found filling pores and crevices of sandstone, limestone, or argillaceous sediments (sediments containing, made of, or resembling clay; clayey), in which case the organic and associated mineral matrix is known as rock asphalt (Speight, 2014 and references cited therein).

Bitumen is a naturally occurring material that is found in deposits where the permeability is low and passage of fluids through the deposit can only be achieved by prior application of fracturing techniques. Tar sand bitumen is a high-boiling material with little, if any, material boiling below 350°C (660°F) and the boiling range may approximate the boiling range of an atmospheric residuum (or in some cases, the boiling range of a vacuum residuum).

For clarification and legal purposes, *tar sands* have been defined in the United States through FE 76-4 (United States Congress, 1976) as:

> ... *the several rock types that contain an extremely viscous hydrocarbon which is not recoverable in its natural state by conventional oil well production methods including currently used enhanced recovery techniques. The hydrocarbon-bearing rocks are variously known as bitumen-rocks oil, impregnated rocks, oil sands, and rock asphalt.*

The recovery of the bitumen depends to a large degree on the composition and construction of the sands. Generally, the bitumen found in tar sand deposits is an extremely viscous material that is *immobile under reservoir conditions* and cannot be recovered through a well by the application of secondary or enhanced recovery techniques.

The expression *tar sand* is commonly used in the petroleum industry to describe sandstone reservoirs that are impregnated with a heavy, viscous black crude oil that cannot be retrieved through a well by conventional production techniques (FE 76-4; United States Congress, 1976). However, the term *tar sand* is actually a misnomer; more correctly, the name *tar* is usually applied to the heavy product remaining after the destructive distillation of coal or other organic matter (Speight, 2013a). Similarly, the terms *oil sand* and *oil shale* are misnomers insofar as these deposits or formations do not contain oil but are oil producing through thermal treatment.

Alternative names, such as *bituminous sand* or *oil sand*, are gradually finding usage, with the former name (bituminous sands) more technically correct. The term *oil sand* is also used in the same way as the term *tar sand*, and these terms are used interchangeably throughout this text.

The bitumen in tar sand formations requires a high degree of thermal stimulation for recovery to the extent that some thermal decomposition may have to be induced. Current recovery operations of bitumen in tar sand formations involve use of a mining technique and non-mining techniques are continually being developed (Speight, 2009, 2014).

Pitch Lake is the name that has been applied to a large surface deposit of bitumen. Guanoco Lake in Venezuela covers more than 1100 acres (445 hectares) and contains an estimated 35 000 000 bbl of bitumen. It was used as a commercial source of asphalt from 1891 to 1935. Smaller deposits occur commonly where Tertiary marine sediments outcrop on the surface; an example is the tar pits at Rancho La Brea in Los Angeles (*brea* and *tar* have been used synonymously with *bitumen*). Although most pitch lakes are fossils of formerly active seeps, some, such as the *Pitch Lake* on the island of Trinidad (also called the *Trinidad Asphalt Lake*), continue to be supplied with fresh crude oil seeping from a subterranean source. The Trinidad *Pitch Lake* covers 115 acres and contains an estimated 40 000 000 bbl of bitumen.

2.6 Biomass

Biomass includes a wide range of materials that produce a variety of products which are dependent upon the feedstock (Balat, 2011; Demirbaş, 2011; Ramroop Singh, 2011; Speight, 2011a). For example, typical biomass wastes include wood material (bark, chips, scraps, and saw dust), pulp and paper industry residues, agricultural residues, organic municipal material, sewage, manure, and food processing byproducts. Agricultural residues such as straws, nut shells, fruit shells, fruit seeds, plant stalks and stover, green leaves, and molasses are potential renewable energy resources. Many developing countries have a wide variety of agricultural residues in ample quantities. Large quantities of agricultural plant residues are produced annually worldwide and are vastly underutilized. When agricultural residues are used as fuel, through direct combustion, only a small percentage of their potential energy is available, due to the inefficiency of burners used. Current disposal methods for these agricultural residues have caused widespread environmental concerns. For example, disposal of rice and wheat straw by open-field burning causes air pollution. In addition, the widely varying heat content of the different types of biomass varies widely and must be taken into consideration when designing any conversion process (Jenkins and Ebeling, 1985).

Biomass is biological material that has come from animal, vegetable, or plant matter and is considered to be *carbon neutral* – while the plant is growing, it uses the sun's energy to absorb the same amount of carbon from the atmosphere as it releases into the atmosphere. By maintaining this closed carbon cycle it is felt, with some mathematical meandering, that there is no overall increase in carbon dioxide levels through emissions to the atmosphere.

Raw materials that can be used to produce biomass fuels are widely available and arise from a large number of different sources and in numerous forms. Biomass can also be used to produce electricity – either blended with traditional feedstocks, such as coal or by itself. However, each of the biomass materials can be used to produce fuel but not all forms are suitable for all the different types of energy conversion technologies such as biomass gasification (Rajvanshi, 1986; Brigwater, 2003; Dasappa et al., 2004; Speight, 2011a; Basu, 2013). The main basic sources of biomass material are: (1) wood, including bark, logs, sawdust, wood chips, wood pellets, and briquettes; (2) high yield energy crops, such as wheat, that are grown specifically for energy applications; (3) agricultural crop and animal residues, like straw or slurry; (4) food waste, both domestic and commercial; and (5) industrial waste, such as wood pulp or paper pulp. For processing, a simple form of biomass such as untreated and unfinished wood may be cut into a number of physical forms, including pellets and wood chips, for use in biomass boilers and stoves.

Thermal conversion processes use heat as the dominant mechanism to convert biomass into another chemical form. The basic alternatives of combustion, torrefaction, pyrolysis, and gasification, are separated principally by the extent to which the chemical reactions involved are allowed to proceed (mainly controlled by the availability of oxygen and conversion temperature) (Speight, 2011a).

Energy created by burning biomass (fuel wood), also known as *dendrothermal energy*, is particularly suited for countries where the fuel wood grow more rapidly, e.g., tropical countries. There is a number of other less common, more experimental or proprietary thermal processes that may offer benefits such as hydrothermal upgrading (HTU) and hydroprocessing. Some have been developed for use on high moisture content biomass, including aqueous slurries, and allow them to be converted into more convenient forms. Some of the applications of thermal conversion are combined heat and power (CHP) and co-firing. In a typical dedicated biomass power plant, efficiencies range from 7 to 27% w/w of the energy content of the fuel. In contrast, co-firing biomass with coal typically occurs at efficiencies near those of the coal combustor (30 to 40% of the energy content of the fuel) (Baxter, 2005; Liu et al., 2011).

Many forms of biomass contain a high percentage of moisture (along with carbohydrates and sugars) and mineral constituents – both of which can influence the viability of a gasification process (Chapter 3) – the presence of high levels of moisture in the biomass reduces the temperature inside the gasifier, which then reduces the efficiency of the gasifier. Therefore, many biomass gasification technologies require that the biomass be dried to reduce the moisture content prior to feeding into the gasifier. In addition, biomass can come in a range of sizes. In many biomass gasification systems, the biomass must be processed to a uniform size or shape to feed into the gasifier at a consistent rate and to ensure that as much of the biomass is gasified as possible.

Biomass, such as wood pellets, yard and crop wastes, and the so-called *energy crops* such as switch grass and waste from pulp and paper mills can be used to produce ethanol and synthetic diesel fuel. The biomass is first gasified to produce the synthetic gas (synthesis gas), and then converted via catalytic processes to these downstream products. Furthermore, most biomass gasification systems use air instead of oxygen for the gasification reactions (which is typically used in large-scale industrial and power gasification plants). Gasifiers that use oxygen require an air separation unit to provide the gaseous/liquid oxygen; this is usually not cost-effective at the smaller scales used in biomass gasification plants. Air-blown gasifiers use the oxygen in the air for the gasification reactions.

In general, biomass gasification plants are much smaller than the typical coal or petroleum coke gasification plants used in the power, chemical, fertilizer, and refining industries – the sustainability of the fuel supply is often brought into question. As such, a biomass gasification plant is less expensive to construct and has a smaller environmental footprint. For example, while a large industrial gasification plant may take up 150 acres of land and process 2500 to 15 000 tons per day of feedstock (such as coal or petroleum coke), the smaller biomass plants typically process 25 to 200 tons of feedstock per day and take up less than 10 acres.

Biomass gasification has been the focus of research in recent years to estimate efficiency and performance of the gasification process using various types of biomass such as sugarcane residue (Gabra et al., 2001), rice hulls (Boateng et al., 1992), pine sawdust (Lv et al., 2004), almond shells (Rapagnà and Latif, 1997; Rapagnà et al.,2000), wheat straw (Ergudenler and Ghaly, 1993), food waste (Ko et al., 2001), and wood biomass (Pakdel and Roy, 1991; Bhattacharya et al., 1999; Chen et al., 1992; Hanaoka et al., 2005). Recently, co-gasification of various biomass and coal mixtures has attracted a great deal of interest from the scientific community. Feedstock combinations including Japanese cedar wood and coal (Kumabe et al., 2007), coal and saw dust (Vélez et al., 2009), coal and pine chips (Pan et al., 2000), coal and silver birch wood (Collot et al., 1999), and coal and birch wood (Brage et al., 2000) have been reported in gasification practice. Co-gasification of coal and biomass has some synergy – the process not only produces a low carbon footprint on the environment, but also improves the H_2/CO ratio in the produced gas which is required for liquid fuel synthesis (Sjöström et al., 1999; Kumabe et al., 2007). In addition, the inorganic matter present in biomass catalyzes the gasification of coal. However, co-gasification processes require custom fittings and optimized processes for the coal and region-specific wood residues.

While co-gasification of coal and biomass is advantageous from a chemical viewpoint, some practical problems are present on upstream, gasification, and downstream processes. On the upstream side, the particle size of the coal and biomass is required to be uniform for optimum gasification. In addition, moisture content and pretreatment (torrefaction) are very important during upstream processing.

While upstream processing is influential from a material handling point of view, the choice of gasifier operation parameters (temperature, gasifying agent, and catalysts) dictate the product gas composition and quality. Biomass decomposition occurs at a lower temperature than coal and therefore different reactors compatible to the feedstock mixture are required (Speight,2011a; Brar et al., 2012; Speight, 2013a, 2013b). Furthermore, feedstock and gasifier type along with operating parameters not only decide product gas composition but also dictate the amount of impurities to be handled downstream.

Downstream processes need to be modified if coal is co-gasified with biomass. Heavy metal and impurities such as sulfur and mercury present in coal can make synthesis gas difficult to use and unhealthy for the environment. Alkali present in biomass can also cause corrosion problems and high temperatures in downstream pipes. An alternative option to downstream gas cleaning would be to process coal to remove mercury and sulfur prior to feeding into the gasifier.

However, first and foremost, coal and biomass require drying and size reduction before they can be fed into a gasifier. Size reduction is needed to obtain appropriate particle sizes; however, drying is required to achieve moisture content suitable for gasification operations. In addition, biomass densification may be conducted to prepare pellets and improve density and material flow in the feeder areas.

It is recommended that biomass moisture content should be less than 15% w/w prior to gasification. High moisture content reduces the temperature achieved in the gasification zone, thus resulting in incomplete gasification. Forest residues or wood has a fiber saturation point at 30 to 31% moisture content (dry basis) (Brar et al., 2012). Compressive and shear strength of the wood increases with decreased moisture content below the fiber saturation point. In such a situation, water is removed from the cell wall leading to shrinkage. The long-chain molecule constituents of the cell wall move closer to each other and bind more tightly. A high level of moisture, usually injected in the form of steam in the gasification zone, favors formation of a water − gas shift reaction that increases hydrogen concentration in the resulting gas.

The torrefaction process is a thermal treatment of biomass in the absence of oxygen, usually at 250 to 300°C (480 to 570°F) to drive off moisture, decompose hemicellulose completely, and partially decompose cellulose (Speight, 2011a). Torrefied biomass has reactive and unstable cellulose molecules with broken hydrogen bonds and not only retains 79 to 95% of feedstock energy but also produces a more reactive feedstock with lower atomic hydrogen − carbon and oxygen − carbon ratios to those of the original biomass. Torrefaction results in higher yields of hydrogen and carbon monoxide in the gasification process.

Most small- to medium-sized biomass/waste gasifiers are air blown, operate at atmospheric pressure, and at temperatures in the range 800 to 100°C (1470 to 2190°F). They face very different challenges to large gasification plants, such as the use of small-scale air separation plant should oxygen gasification be preferred. Pressurized operation, which eases gas cleaning, may not be practical.

Biomass fuel producers, coal producers, and, to a lesser extent, waste companies are enthusiastic about supplying co-gasification power plants and realize the benefits of co-gasification with alternate fuels (Speight, 2008, 2011a; Lee and Shah, 2013; Speight, 2013a, 2013b). The benefits of a co-gasification technology involving coal and biomass include the use of a reliable coal supply with gate-fee waste and biomass which allows the economies of scale from a larger plant to be supplied just with waste and biomass. In addition, the technology offers a future option of hydrogen production and fuel development in refineries. In fact, oil refineries and petrochemical plants are opportunities for gasifiers when the hydrogen is particularly valuable (Speight, 2011b, 2014).

In addition, while biomass may seem to some observers to be the answer to the global climate change issue, the advantages and disadvantages must be considered carefully. For example, the advantages are: (1) biomass is a theoretically inexhaustible fuel source; (2) when direct conversion of combustion of plant mass − such as fermentation and pyrolysis − is not used to generate energy there is minimal environmental impact; (3) alcohols and other fuels produced by biomass are efficient, viable, and relatively clean-burning; and (4) biomass is available on a worldwide basis. On the other hand, the disadvantages include (1) the highly variable heat content of different biomass feedstocks, (2) the high water content that can affect the process of energy balance, and (3) there is a potential net loss of energy when a biomass plant is operated on a small scale − an account of the energy output used to grow and harvest the biomass must be included in the energy balance.

2.7 Solid Waste

Waste may be municipal solid waste (MSW) which had minimal pre-sorting, or refuse-derived fuel (RDF) with significant pretreatment, usually mechanical screening and shredding. Other more specific waste sources (excluding hazardous waste) and possibly including petroleum coke may provide niche opportunities for co-utilization (Brigwater, 2003; Arena, 2012; Basu, 2013; Speight, 2013a, 2013b).

The traditional waste-to-energy plant, based on mass-burn combustion on an inclined grate, has a low public acceptability despite the very low emissions achieved over the last decade with modern flue gas cleanup equipment. This has led to difficulty in obtaining planning permissions to construct needed new waste-to-energy plants. After much debate, various governments have allowed options for advanced waste conversion technologies (gasification, pyrolysis, and anaerobic digestion), but will only give credit to the proportion of electricity generated from non-fossil waste.

Use of waste materials as co-gasification feedstocks may attract significant disposal credits (Ricketts et al., 2002). Cleaner biomass materials are renewable fuels and may attract premium prices for the electricity generated. Availability of sufficient fuel locally for an economic plant size is often a major issue, as is the reliability of the fuel supply. Use of more predictably available coal alongside these fuels overcomes some of these difficulties and risks. Coal could be regarded as the base feedstock which keeps the plant running when the fuels producing the better revenue streams are not available in sufficient quantities.

Coal characteristics are very different to younger hydrocarbon fuels such as biomass and waste. Hydrogen-to-carbon ratios are higher for younger fuels, as is the oxygen content. This means that reactivity is very different under gasification conditions. Gas cleaning issues can also be very different, sulfur being a major concern for coal gasification and chlorine compounds and tars more important for waste and biomass gasification. There are no current proposals for adjacent gasifiers and gas cleaning systems, one handling biomass or waste and one coal, alongside each other and feeding the same power production equipment. However, there are some advantages to such a design as compared with mixing fuels in the same gasifier and gas cleaning systems.

Electricity production or combined electricity and heat production remain the most likely area for the application of gasification or co-gasification. The lowest investment cost per unit of electricity generated is the use of the gas in an existing large power station. This has been done in several large utility boilers, often with the gas fired alongside the main fuel. This option allows a comparatively small thermal output of gas to be used with the same efficiency as the main fuel in the boiler as a large, efficient steam turbine can be used. It is anticipated that addition of gas from a biomass or wood gasifier into the natural gas feed to a gas turbine is technically possible but there will be concerns as to the balance of commercial risks to a large power plant and the benefits of using the gas from the gasifier.

Furthermore, the disposal of municipal and industrial waste has become an important problem because the traditional means of disposal, landfill, are much less environmentally acceptable than previously. Much stricter regulation of these disposal methods will make the economics of waste processing for resource recovery much more favorable.

One method of processing waste streams is to convert the energy value of the combustible waste into a fuel. One type of fuel attainable from waste is a low heating value gas, usually $100 - 150$ Btu/scf, which can be used to generate process steam or to generate electricity (Gay et al., 1980). Co-processing such waste with coal is also an option (Speight, 2008).

Co-gasification technology varies, usually being site specific and high feedstock dependent. At the largest scale, the plant may include the well proven fixed bed and entrained flow gasification processes. At smaller scales, emphasis is placed on technologies which appear closest to commercial operation. Pyrolysis and other advanced thermal conversion processes are included where power generation is practical using the on-site feedstock produced. However, the needs to be addressed are (1) core fuel handling and gasification/pyrolysis technologies, (2) fuel gas cleanup, and (3) conversion of fuel gas to electric power (Ricketts et al., 2002).

The traditional waste-to-energy plant, based on mass-burn combustion on an inclined grate, has a low public acceptability despite the very low emissions achieved over the last decade with modern flue gas cleanup equipment. This has led to difficulty in obtaining planning permission to construct needed new waste-to-energy plants. After much debate, various governments have allowed options for advanced waste conversion technologies (gasification, pyrolysis, and anaerobic digestion), but will only give credit to the proportion of electricity generated from non-fossil waste.

Co-utilization of waste and biomass with coal may provide economies of scale that help achieve the above identified policy objectives at an affordable cost. In some countries, governments propose co-gasification processes as being *well suited for community-sized developments* suggesting that waste should be dealt with in smaller plants serving towns and cities, rather than moved to large, central plants (satisfying the so-called *proximity principal*).

In fact, neither biomass nor wastes are currently produced, or naturally gathered at sites in sufficient quantities to fuel a modern, large, and efficient power plant. Disruption, transport issues, fuel use, and public opinion all act against gathering hundreds of megawatts (MWe) at a single location. Biomass or waste-fired power plants are therefore inherently limited in size and hence in efficiency (labor costs per unit electricity produced) and in other economies of scale. The production rates of municipal refuse follow reasonably predictable patterns over time periods of a few years. Recent experience with the very limited current *biomass for energy* harvesting has shown unpredictable variations in harvesting capability with long periods of zero production over large areas during wet weather.

The situation is very different for coal. This is generally mined or imported and thus large quantities are available from a single source or a number of closely located sources, and supply has been reliable and predictable. However, the economics of new coal-fired power plants of any technology or size have not encouraged any new coal-fired power plant in the gas generation market.

The potential unreliability of biomass, longer-term changes in refuse, and the size limitation of a power plant using only waste and/or biomass can be overcome combining biomass, refuse, and coal. It also allows benefit from a premium electricity price for electricity from biomass and the gate fee associated with waste. If the power plant is gasification-based, rather than direct combustion, further benefits may be available.

These include a premium price for the electricity from waste, the range of technologies available for the gas to electricity part of the process, gas cleaning prior to the main combustion stage instead of after combustion, and public image, which is currently generally better for gasification than for combustion. These considerations have led to the current studies of co-gasification of wastes/biomass with coal (Speight, 2008).

For large-scale power generation ($>$50 MWe), the gasification field is dominated by plant based on the pressurized, oxygen-blown, entrained flow or fixed-bed gasification of fossil fuels. Entrained gasifier operational experience to date has largely been with well-controlled fuel feedstocks with short-term trial work at low co-gasification ratios and with easily handled fuels.

Use of waste materials as co-gasification feedstocks may attract significant disposal credits. Cleaner biomass materials are renewable fuels and may attract premium prices for the electricity generated. Availability of sufficient fuel locally for an economic plant size is often a major issue, as is the reliability of the fuel supply. Use of more-predictably available coal alongside these fuels overcomes some of these difficulties and risks. Coal could be regarded as the *flywheel* which keeps the plant running when the fuels producing the better revenue streams are not available in sufficient quantities.

Electricity production or combined electricity and heat production remain the most likely area for the application of gasification or co-gasification. The lowest investment cost per unit of electricity generated is the use of the gas in an existing large power station. This has been done in several large utility boilers, often with the gas fired alongside the main fuel. This option allows a comparatively small thermal output of gas to be used with the same efficiency as the main fuel in the boiler as a large, efficient steam turbine can be used. It is anticipated that addition of gas from a biomass or wood gasifier into the natural gas feed to a gas turbine is technically possible but there will be concerns as to the balance of commercial risks to a large power plant and the benefits of using the gas from the gasifier.

Furthermore, the disposal of municipal and industrial waste has become an important problem because the traditional means of disposal, landfill, are much less environmentally acceptable than previously.

Much stricter regulation of these disposal methods will make the economics of waste processing for resource recovery much more favorable.

One method of processing waste streams is to convert the energy value of the combustible waste into a fuel. One type of fuel attainable from waste is a low heating value gas, usually $100 - 150$ Btu/scf, which can be used to generate process steam or to generate electricity (Gay et al., 1980). Co-processing such waste with coal is also an option (Speight, 2008, 2013a, 2013b).

In summary, coal may be co-gasified with waste or biomass for environmental, technical, or commercial reasons. It allows larger, more efficient plants than those sized for grown biomass or arising waste within a reasonable transport distance; specific operating costs are likely to be lower and fuel supply security is assured.

Co-gasification technology varies, being usually site specific and high feedstock dependent. At the largest scale, the plant may include the well proven fixed bed and entrained flow gasification processes. At smaller scales, emphasis is placed on technologies which appear closest to commercial operation. Pyrolysis and other advanced thermal conversion processes are included where power generation is practical using the on-site feedstock produced. However, the needs to be addressed are: (1) core fuel handling and gasification/pyrolysis technologies, (2) fuel gas cleanup, and (3) conversion of fuel gas to electric power (Ricketts et al., 2002).

Co-utilization of waste and biomass with coal may provide economies of scale that help achieve the above identified policy objectives at an affordable cost. In some countries, governments propose co-gasification processes as being *well suited for community-sized developments* suggesting that waste should be dealt with in smaller plants serving towns and cities, rather than moved to large, central plants (satisfying the so-called *proximity principle*).

In fact, neither biomass nor wastes are currently produced or naturally gathered at sites in sufficient quantities to fuel a modern large and efficient power plant. Disruption, transport issues, fuel use, and public opinion all act against gathering hundreds of megawatts (MWe) at a single location. Biomass or waste-fired power plants are therefore inherently limited in size and hence in efficiency (labor costs per unit electricity produced) and in other economies of scale. The production rates of municipal refuse follow reasonably predictable patterns over time periods of a few years. Recent experience with the very limited current *biomass for energy* harvesting has shown unpredictable variations in harvesting capability with long periods of zero production over large areas during wet weather.

The situation is very different for coal. This is generally mined or imported and thus large quantities are available from a single source or a number of closely located sources, and supply has been reliable and predictable. However, the economics of new coal-fired power plants of any technology or size have not encouraged any new coal-fired power plant in the gas generation market.

The potential unreliability of biomass, longer-term changes in refuse, and the size limitation of a power plant using only waste and/or biomass can be overcome combining biomass, refuse, and coal. It also allows benefit from a premium electricity price for electricity from biomass and the gate fee associated with waste. If the power plant is gasification-based, rather than direct combustion, further benefits may be available. These include a premium price for the electricity from waste, the range of technologies available for the gas to electricity part of the process, gas cleaning prior to the main combustion stage instead of after combustion and public image, which is currently generally better for gasification as compared to combustion. These considerations lead to current studies of co-gasification of wastes/biomass with coal (Speight, 2008).

For large-scale power generation (>50 MWe), the gasification field is dominated by plant based on the pressurized, oxygen-blown, entrained flow or fixed-bed gasification of fossil fuels. Entrained gasifier operational experience to date has largely been with well-controlled fuel feedstocks with short-term trial work at low co-gasification ratios and with easily handled fuels.

Analyses of the composition of municipal solid waste indicate that plastics do make up measureable amounts (5 to 10% or more) of solid waste streams (EPCI, 2004; Mastellone and Arena, 2007). Many of these plastics are worth recovering as energy. In fact, many plastics, particularly the poly-olefins, have high calorific values and simple chemical constitutions of primarily carbon and hydrogen. As a result, waste plastics are ideal candidates for the gasification process. Because of the myriad of sizes and shapes of plastic products size reduction is necessary to create a feed material of a size less than 2 inches in diameter. Some forms of waste plastics such as thin films may require a simple agglomeration step to produce a particle of higher bulk density to facilitate ease of feeding. A plastic, such as high-density polyethylene, processed through a gasifier is converted to carbon monoxide and hydrogen and these materials in turn may be used to form other chemicals including ethylene from which the polyethylene is produced − *closed loop recycling*.

2.8 Black Liquor

Black liquor is the spent liquor from the Kraft process in which pulpwood is converted into paper pulp by removing lignin constituents, hemicellulose constituents, and other extractable materials from wood to free the cellulose fibers. The equivalent spent cooking liquor in the sulfite process is usually called *brown liquor*, but the terms *red liquor*, *thick liquor*, and *sulfite liquor* are also used. Approximately seven units of black liquor are produced in the manufacture of one unit of pulp (Biermann, 1993).

Black liquor is an aqueous solution of lignin residues, hemicellulose, and the inorganic chemical used in the process and 15% w/w solids of which 10% w/w are inorganic and 5% w/w are organic. Typically, the organic constituents in black liquor are 40 to 45% w/w soaps, 35 to 45% w/w lignin, and 10 to 15% w/w other (miscellaneous) organic materials.

The organic constituents in the black liquor are made up of water/alkali soluble degradation components from the wood. Lignin is degraded to shorter fragments with sulfur content on the order of 1 to 2% w/w and sodium content at approximately 6% w/w of the dry solids. Cellulose (and hemicellulose) is degraded to aliphatic carboxylic acid soaps and hemicellulose fragments. The extractable constituents yield *tall oil soap* and crude turpentine. The tall oil soap may contain up to 20% w/w sodium. The residual lignin components currently serve for hydrolytic or pyrolytic conversion or combustion. Alternatively hemicellulose constituents may be used in fermentation processes.

Black liquor gasification has the potential to achieve higher overall energy efficiency than the conventional recovery boiler while generating an energy-rich *synthesis gas* from the liquor. The synthesis gas can be burned in a gas turbine combined cycle system (BLGCC — black liquor gasification combined cycle — and similar to IGCC, integrated gasification combined cycle) to produce electricity or converted (through catalytic processes) into chemicals or fuels such as methanol, dimethyl ether, and Fischer — Tropsch hydrocarbons, such as diesel fuel.

REFERENCES

Arena, U., 2012. Process and technological aspects of municipal solid waste gasification. Rev. Waste Manage. 32, 625–639.

Balat, M., 2011. Fuels from biomass — an overview. In: Speight, J.G. (Ed.), The Biofuels Handbook. Royal Society of Chemistry, London, UK, Part 1, Chapter 3.

Basu, P., 2013. Biomass Gasification, Pyrolysis and Torrefaction 2nd Edition: Practical Design and Theory. Academic Press, Inc., New York.

Baxter, L., 2005. Biomass-coal co-combustion: opportunity for affordable renewable energy. Fuel 84 (10), 1295–1302.

Bhattacharya, S., Md. Mizanur Rahman Siddique, A.H., Pham, H-L., 1999. A study in wood gasification on low tar production. Energy 24, 285–296.

Biermann, C.J., 1993. Essentials of Pulping and Papermaking. Academic Press Inc., New York.

Boateng, A.A., Walawender, W.P., Fan, L.T., Chee, C.S., 1992. Fluidized-Bed steam gasification of rice hull. Bioresour. Technol. 40 (3), 235–239.

Brage, C., Yu, Q., Chen, G., Sjöström, K., 2000. Tar evolution profiles obtained from gasification of biomass and coal. Biomass Bioenergy 18 (1), 87–91.

Brar, J.S., Singh, K., Wang, J., Kumar, S., 2012. Cogasification of coal and biomass: a review. Int. J. Forestry Res. 2012, 1–10.

Brigwater, A.V. (Ed.), 2003. Pyrolysis and Gasification of Biomass and Waste. CPL Press, Newbury, Berkshire, UK.

Chadeesingh, R., 2011. The fischer – tropsch process. In: Speight, J.G. (Ed.), The Biofuels Handbook. Royal Society of Chemistry, London, UK, pp. 476–517. (Part 3, Chapter 5).

Chen, G., Sjöström, K., Bjornbom, E., 1992. Pyrolysis/Gasification of wood in a pressurized fluidized bed reactor. Ind. Eng. Chem. Res. 31 (12), 2764–2768.

Collot, A.G., Zhuo, Y., Dugwell, D.R., Kandiyoti, R., 1999. Co-Pyrolysis and cogasification of coal and biomass in bench-scale fixed-bed and fluidized bed reactors. Fuel 78, 667–679.

Dasappa, S., Paul, P.J., Mukunda, H.S., Rajan, N.K.S., Sridhar, G., Sridhar, H.V., 2004. Biomass gasification technology — a route to meet energy needs. Curr. Sci. 87 (7), 908–916.

Demirbaş, A., 2005. Fuel and combustion properties of bio-wastes. Energy Sources Part A 27, 451–462.

Demirbaş, A., 2011. Production of fuels from crops. In: Speight, J.G. (Ed.), The Biofuels Handbook. Royal Society of Chemistry, London, UK (Part 2, Chapter 1).

Dry, M.E., 1976. Advances in Fischer – Tropsch Chemistry. Ind. Eng. Chem. Res. 15 (4), 282–286.

EIA, 2007. Net Generation by Energy Source by Type of Producer. Energy Information Administration, United States Department of Energy, Washington, DC. <http://www.eia.doe.gov/cneaf/electricity/epm/table1_1.html>.

EPCI, 2004. The Gasification of Residual Plastics Derived from Municipal Recycling Facilities. The Environment and Industry Council, Canadian Plastics Industry Association, Ottawa, Ontario, Canada.

Ergudenler, A., Ghaly, A.E., 1993. Agglomeration of alumina sand in a fluidized bed straw gasifier at elevated temperatures. Bioresour. Technol. 43 (3), 259–268.

Gabra, M., Pettersson, E., Backman, R., Kjellström, B., 2001. Evaluation of cyclone gasifier performance for gasification of sugar cane residue – part 1: gasification of bagasse. Biomass Bioenergy 21 (5), 351–369.

Gay, R.L., Barclay, K.M., Grantham, L.F., Yosim, S.J., 1980. Fuel production from solid waste. Symposium on Thermal Conversion of Solid Waste and Biomass. Symposium Series No. 130, American Chemical Society, Washington, DC. Chapter 17, pp. 227–236.

Hanaoka, T., Inoue, S., Uno, S., Ogi, T., Minowa, T., 2005. Effect of woody biomass components on air-steam gasification. Biomass Bioenergy 28 (1), 69–76.

Jenkins, B.M., Ebeling, J.M., 1985. Thermochemical Properties of Biomass Fuels. Calif. Agric.14–18, May–June.

Ko, M.K., Lee, W.Y., Kim, S.B., Lee, K.W., Chun, H.S., 2001. Gasification of food waste with steam in fluidized bed. Korean J. Chem. Eng. 18 (6), 961–964.

Kumabe, K., Hanaoka, T., Fujimoto, S., Minowa, T., Sakanishi, K., 2007. Cogasification of woody biomass and coal with air and steam. Fuel 86, 684–689.

Lee, S., Shah, Y.T., 2013. Biofuels and Bioenergy. CRC Press, Taylor & Francis Group, Boca Raton, FL.

Lee, S., Speight, J.G., Loyalka, S., 2007. Handbook of Alternative Fuel Technologies. CRC-Taylor & Francis Group, Boca Raton, FL.

Liu, G., Larson, E.D., Williams, R.H., Kreutz, T.G., Guo, X., 2011. Making fischer – tropsch fuels and electricity from coal and biomass: performance and cost analysis. Energy Fuels 25, 415–437.

Lv, P.M., Xiong, Z.H., Chang, J., Wu, C.Z., Chen, Y., Zhu, J.X., 2004. An experimental study on biomass air-steam gasification in a fluidized bed. Bioresour. Technol. 95 (1), 95–101.

Mastellone, M.L., Arena, U., 2007. Fluidized bed gasification of plastic waste: effect of bed material on process performance. Proceedings. 55th International Energy Agency – Fluidized Bed Conversion Meeting, Paris, France, October 30 – 31.

Mukoma, P., Hildebrandt, D., David Glasser, D., 2006. A process synthesis approach to investigate the effect of the probability of chain growth on the efficiency of fischer–tropsch synthesis. Ind. Eng. Chem. Res. 45, 5928–5935.

Pakdel, H., Roy, C., 1991. Hydrocarbon content of liquid products and tar from pyrolysis and gasification of wood. Energy Fuels 5, 427–436.

Pan, Y.G., Velo, E., Roca, X., Manyà, J.J., Puigjaner, L., 2000. Fluidized-Bed cogasification of residual biomass/poor coal blends for fuel gas production. Fuel 79, 1317–1326.

Rajvanshi, A.K., 1986. Biomass gasification. In: Goswami, D.Y. (Ed.), Alternative Energy in Agriculture, vol. II. CRC Press, Boca Raton, FL, pp. 83–102.

Ramroop Singh, N., 2011. Biofuel. In: Speight, J.G. (Ed.), The Biofuels Handbook. Royal Society of Chemistry, London, UK (Part 1, Chapter 5).

Rapagnà, N.J., Latif, A., 1997. Steam gasification of almond shells in a fluidized bed reactor: the influence of temperature and particle size on product yield and distribution. Biomass Bioenergy 12 (4), 281–288.

Rapagnà, N.J., Kiennemann, A., Foscolo, P.U., 2000. Steam-gasification of biomass in a fluidized-bed of olivine particles. Biomass Bioenergy 19 (3), 187–197.

Ricketts, B., Hotchkiss, R., Livingston, W., Hall, M., 2002. Technology status review of waste/biomass co-gasification with coal. Proceedings of the Institute of Chemical Engineering Fifth European Gasification Conference. Noordwijk, Netherlands, April 8–10.

Sjöström, K., Chen, G., Yu, Q., Brage, C., Rosén, C., 1999. Promoted reactivity of char in cogasification of biomass and coal: synergies in the thermochemical process. Fuel 78, 1189–1194.

Speight, J.G., 2008. Synthetic Fuels Handbook: Properties, Processes, and Performance. McGraw-Hill, New York.

Speight, J.G., 2009. Enhanced Recovery Methods for Heavy Oil and Tar Sands. Gulf Publishing Company, Houston, Texas.

Speight, J.G. (Ed.), 2011a. The Biofuels Handbook. Royal Society of Chemistry, London, UK.

Speight, J.G., 2011b. The Refinery of the Future. Gulf Professional Publishing, Elsevier, Oxford, UK.

Speight, J.G., 2013a. The Chemistry and Technology of Coal 3rd Edition. CRC Press, Taylor & Francis Group, Boca Raton, FL.

Speight, J.G., 2013b. Coal-Fired Power Generation Handbook. Scrivener Publishing, Salem, MA.

Speight, J.G., 2013c. Oil Sand Production Processes. Gulf Professional Publishing, Elsevier, Oxford, UK.

Speight, J.G., 2014. The Chemistry and Technology of Petroleum 5th Edition. CRC Press, Taylor & Francis Group, Boca Raton, FL.

US Congress., 1976. Public Law FEA-76-4. United States Library of Congress, Washington, DC.

Vélez, F.F., Chejne, F., Valdés, C.F., Emery, E.J., Londoño, C.A., 2009. Cogasification of colombian coal and biomass in a fluidized bed: an experimental study. Fuel 88 (3), 424–430.

CHAPTER 2

Chemistry of Gasification

1 INTRODUCTION

The gasification of any carbonaceous or hydrocarbonaceous material is, essentially, the conversion of the carbon constituents by any one of a variety of chemical processes to produce combustible gases (Higman and van der Burgt, 2008; Speight, 2008, 2013a). With the rapid increase in the use of gasification technology from the 19th century onwards it is not surprising the concept of producing a flammable gas for domestic heating, industrial heating, and power generation became common-place in the 19th and 20th centuries (Speight, 2013a, 2013b).

Gasification includes a series of reaction steps that convert the feedstock into *synthesis gas* (carbon monoxide, CO, plus hydrogen, H_2) and other gaseous products. This conversion is generally accomplished by introducing a gasifying agent (air, oxygen, and/or steam) into a reactor vessel containing the feedstock where the temperature, pressure, and flow pattern (moving bed, fluidized, or entrained bed) are controlled. The gaseous products — other than carbon monoxide and hydrogen — and the proportions of these product gases (such as carbon dioxide, CO_2, methane, CH_4, water vapor, H_2O, hydrogen sulfide, H_2S, and sulfur dioxide, SO_2) depend on: the (1) type of feedstock, (2) the chemical composition of the feedstock, (3) the gasifying agent or gasifying medium, as well as (4) the thermodynamics and chemistry of the gasification reactions as controlled by the process operating parameters (Singh et al., 1980; Pepiot et al., 2010; Shabbar and Janajreh, 2013; Speight, 2013a, 2013b). In addition, the kinetic rates and extents of conversion for the several chemical reactions that are a part of the gasification process are variable and are typically functions of: (1) temperature, (2) pressure, (3) reactor configuration, and (4) the gas composition of the product gases and whether or not these gases influence the outcome of the reaction (Johnson, 1979; Penner, 1987; Müller et al., 2003; Slavinskaya et al., 2009; Speight, 2013a, 2013b).

Generally, the reaction rate (i.e., the rate of feedstock conversion) is higher at higher temperatures, whereas reaction equilibrium may be favored at either higher or lower temperatures depending on the specific type of gasification reaction. The effect of pressure on the rate also depends on the specific reaction. Thermodynamically, some gasification reactions such as the carbon − hydrogen reaction to produce methane are favored at high pressures (>1030 psi) and relatively lower temperatures (760 to 930°C; 1400 to 1705°F), whereas low pressures and high temperatures favor the production of synthesis gas (i.e., carbon monoxide and hydrogen) via the steam or carbon dioxide gasification reaction.

Because of the overall complexity of the gasification process, it is necessary to present a description of the chemistry of the gasification reactions and it is the purpose of this chapter to present descriptions of the various reactions involved in gasification of carbonaceous and hydrocarbonaceous feedstocks as well as the various thermodynamic aspects of these reactions which dictate the process parameters used to produce the various gases.

2 CHEMICAL CONCEPTS

Chemically, gasification involves the thermal decomposition of the feedstock and the reaction of the feedstock carbon and other pyrolysis products with oxygen, water, and fuel gases such as methane (Table 2.1). In fact, gasification is often considered to involve two

Table 2.1 Gasification Reactions
$2C + O_2 \rightarrow 2CO$
$C + O_2 \rightarrow CO_2$
$C + CO_2 \rightarrow 2CO$
$CO + H_2O \rightarrow CO_2 + H_2$ (shift reaction)
$C + H_2O \rightarrow CO_2 + H_2$ (water gas reaction)
$C + 2H_2 \rightarrow CH_4$
$2H_2 + O_2 \rightarrow + 2H_2O$
$CO + 2H_2 \rightarrow CH_3OH$
$CO + 3H_2 \rightarrow CH_4 + H_2O$ (methanation reaction)
$CO_2 + 4H_2 \rightarrow CH_4 + 2H_2O$
$C + 2H_2O \rightarrow 2H_2 + CO_2$
$2C + H_2 \rightarrow C_2H_2$
$CH_4 + 2H_2O \rightarrow CO_2 + 4H_2$

distinct chemical stages: (1) devolatilization of the feedstock to produce volatile matter and char, followed by (2) char gasification, which is complex and specific to the conditions of the reaction – both processes contribute to the complex kinetics of the gasification process (Sundaresan and Amundson, 1978).

Gasification of char in an atmosphere of carbon dioxide can be divided into two stages: (1) pyrolysis and (2) gasification of the pyrolytic char. In the first stage, pyrolysis (removal of moisture content and devolatilization) occurs at a comparatively lower temperature. In the second stage, gasification of the pyrolytic char is achieved by reaction with oxygen/carbon dioxide mixtures at high temperature. In nitrogen and carbon dioxide environments from room temperature to 1000°C (1830°F), the mass loss rate of pyrolysis in nitrogen may be significantly different (sometimes lower, depending on the feedstock) from mass loss rate in carbon dioxide, which may be due (in part) to the difference in properties of the bulk gases.

Using coal as an example, gasification in an atmosphere of oxygen/carbon dioxide is almost the same as gasification in an atmosphere of oxygen/nitrogen at the same oxygen concentration but this effect is slightly delayed at high temperature. This may be due to the lower rate of diffusion of oxygen through carbon dioxide and the higher specific heat capacity of carbon dioxide. However with an increase in the concentration of oxygen, the mass loss rate of coal also increases and, hence, shortens the burn out time of coal. The optimum value of oxygen/carbon dioxide for the reaction of oxygen with the functional groups that are present in the coal feedstock is on the order of 8% v/v.

2.1 General Aspects

In a gasifier, the feedstock particle is exposed to high temperatures generated from the partial oxidation of the carbon. As the particle is heated, any residual moisture (assuming that the feedstock has been pre-fired) is driven off and further heating of the particle begins to drive off the volatile gases. Discharge of the volatile products will generate a wide spectrum of hydrocarbons ranging from carbon monoxide and methane to long-chain hydrocarbons comprising tars, creosote, and heavy oil. The complexity of the products will also affect the progress and rate of the reaction, each product being produced by a different chemical process at a different rate. At a temperature above 500°C

(930°F) the conversion of the feedstock to char and ash and char is completed. In most of the early gasification processes, this was the desired byproduct but for gas generation the char provides the necessary energy to effect further heating and, typically, the char is contacted with air or oxygen and steam to generate the product gases.

Furthermore, with an increase in heating rate, feedstock particles are heated more rapidly and are burned in a higher temperature region, but the increase in heating rate has almost no effect on the mechanism (Irfan, 2009).

Most notable effects in the physical chemistry of the gasification process are those due to the chemical character of the feedstock as well as the physical composition of the feedstock (Speight, 2011a, 2013b). In more general terms of the character of the feedstock, gasification technologies generally require some initial processing of the feedstock with the type and degree of pretreatment a function of the process and/ or the type of feedstock. For example, the Lurgi process will accept lump feedstock (1 inch, 25 mm, to 28 mesh), but it must be non-caking with the fines removed − feedstock that shows caking or agglomerating tendencies form a plastic mass in the bottom of a gasifier and subsequently plug up the system, thereby markedly reducing process efficiency. Thus, some attempt to reduce caking tendencies is necessary and can involve preliminary partial oxidation of the feedstock thereby destroying the caking properties.

Another factor, often presented as a very general *rule of thumb*, is that optimum gas yields and gas quality are obtained at operating temperatures of approximately 595 to 650°C (1100 to 1200°F. A gaseous product with a higher heat content (BTU/ft^3) can be obtained at lower system temperatures but the overall yield of gas (determined as the *fuel-to-gas ratio*) is reduced by the unburned char fraction.

With some feedstocks, the higher the amounts of volatile material produced in the early stages of the process the higher the heat content of the product gas. In some cases, the highest gas quality may be produced at the lowest temperatures but when the temperature is too low, char oxidation reaction is suppressed and the overall heat content of the product gas is diminished. All such events serve to complicate the reaction rate and make derivatives of a global kinetic relationship applicable to all types of feedstock subject to serious question and doubt.

Depending on the type of feedstock being processed and the analysis of the gas product desired, pressure also plays a role in product definition. In fact, some (or all) of the following processing steps will be required: (1) pretreatment of the feedstock; (2) primary gasification of the feedstock; (3) secondary gasification of the carbonaceous residue from the primary gasifier; (4) removal of carbon dioxide, hydrogen sulfide, and other acid gases; (5) shift conversion for adjustment of the carbon monoxide/hydrogen mole ratio to the desired ratio; and (6) catalytic methanation of the carbon monoxide/hydrogen mixture to form methane. If high-heat-content (high-Btu) gas is desired, all of these processing steps are required since gasifiers do not typically yield methane in the significant concentration.

2.2 Pretreatment

While feedstock pretreatment for introduction into the gasifier is often considered to be a physical process in which the feedstock is prepared for gasifier – typically as pellets or finely-ground feedstock – there are chemical aspects that must also be considered.

Some feedstocks, especially certain types of coal, display caking or agglomerating characteristics when heated (Speight, 2013a) and these coal types are usually not amenable to treatment by gasification processes employing fluidized bed or moving bed reactors; in fact, caked coal is difficult to handle in fixed-bed reactors. The pretreatment involves a mild oxidation treatment which destroys the caking characteristics of coals and usually consists of low-temperature heating of the coal in the presence of air or oxygen.

While this may seemingly be applicable to coal gasification only, this form of coal pretreatment is particularly important when a non-coal feedstock is co-gasified with coal.

2.3 Reactions

Gasification involves the thermal decomposition of feedstock and the reaction of the feedstock carbon and other pyrolysis products with oxygen, water, and fuel gases such as methane. The presence of oxygen, hydrogen, water vapor, carbon oxides, and other compounds in the reaction atmosphere during pyrolysis may either support or inhibit numerous reactions with carbonaceous feedstocks and with the products evolved. The distribution of weight and chemical composition of the products are also influenced by the prevailing conditions (i.e., temperature, heating rate, pressure, and residence time) and, last but by no means least, the nature of the feedstock (Wang and Mark, 1992).

If a high-heat-content (high-Btu) gas (900 to 1000 Btu/ft^3) is required, efforts must be made to increase the methane content of the gas. The reactions which generate methane are all exothermic and have negative values, but the reaction rates are relatively slow and catalysts may, therefore, be necessary to accelerate the reactions to acceptable commercial rates. Indeed, the overall reactivity of the feedstock and char may be subject to catalytic effects. It is also possible that the mineral constituents of the feedstock (such as the mineral matter in coal and biomass) may modify the reactivity by a direct catalytic effect (Cusumano et al., 1978; Davidson, 1983; Baker and Rodriguez, 1990; Mims, 1991; Martinez-Alonso and Tascon, 1991).

In the process, the feedstock undergoes three processes in its conversation to synthesis gas – the first two processes, pyrolysis and combustion, occur very rapidly. In pyrolysis, char is produced as the feedstock heats up and volatiles are released. In the combustion process, the volatile products and some of the char reacts with oxygen to produce various products (primarily carbon dioxide and carbon monoxide) and the heat required for subsequent gasification reactions. Finally, in the gasification process, the feedstock char reacts with steam to produce hydrogen (H_2) and carbon monoxide (CO).

Combustion:

$$2C_{feedstock} + O_2 \rightarrow 2CO + H_2O$$

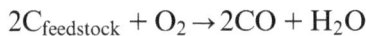

Gasification:

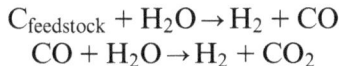

$$C_{feedstock} + H_2O \rightarrow H_2 + CO$$
$$CO + H_2O \rightarrow H_2 + CO_2$$

The resulting synthesis gas is approximately 63% v/v carbon monoxide, 34% v/v hydrogen, and 3% v/v carbon dioxide. At the gasifier temperature, the ash and other feedstock mineral matter liquefies and exits at the bottom of the gasifier as slag, a sand-like inert material that can be sold as a co-product to other industries (e.g., road building). The synthesis gas exits the gasifier at pressure and high temperature and must be cooled prior to the synthesis gas cleaning stage.

Although processes that use the high temperature to raise high-pressure steam are more efficient for electricity production, full-quench cooling, by which the synthesis gas is cooled by the direct injection of water, is more appropriate for hydrogen production. Full-quench cooling provides the necessary steam to facilitate the water − gas shift reaction, in which carbon monoxide is converted to hydrogen and carbon dioxide in the presence of a catalyst:

Water − gas shift reaction:

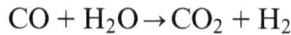

$$CO + H_2O \rightarrow CO_2 + H_2$$

This reaction maximizes the hydrogen content of the synthesis gas, which consists primarily of hydrogen and carbon dioxide at this stage. The synthesis gas is then scrubbed of particulate matter and sulfur is removed via physical absorption (Speight, 2013a, 2014). The carbon dioxide is captured by physical absorption or a membrane and either vented or sequestered.

Thus, in the initial stages of gasification, the rising temperature of the feedstock initiates devolatilization and the breaking of weaker chemical bonds to yield volatile tar, volatile oil, phenol derivatives, and hydrocarbon gases. These products generally react further in the gaseous phase to form hydrogen, carbon monoxide, and carbon dioxide. The char (fixed carbon) that remains after devolatilization reacts with oxygen, steam, carbon dioxide, and hydrogen. Overall, the chemistry of gasification is complex but can be conveniently (and simply) represented by the following reactions:

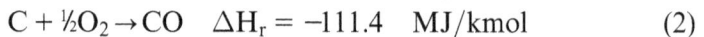

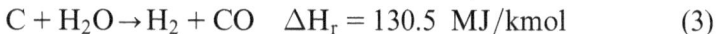

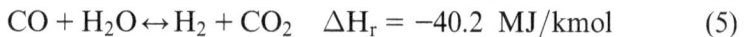

$$C + O_2 \rightarrow CO_2 \quad \Delta H_r = -393.4 \ \text{MJ/kmol} \tag{1}$$
$$C + \tfrac{1}{2}O_2 \rightarrow CO \quad \Delta H_r = -111.4 \ \text{MJ/kmol} \tag{2}$$
$$C + H_2O \rightarrow H_2 + CO \quad \Delta H_r = 130.5 \ \text{MJ/kmol} \tag{3}$$
$$C + CO_2 \leftrightarrow 2CO \quad \Delta H_r = 170.7 \ \text{MJ/kmol} \tag{4}$$
$$CO + H_2O \leftrightarrow H_2 + CO_2 \quad \Delta H_r = -40.2 \ \text{MJ/kmol} \tag{5}$$
$$C + 2H_2 \rightarrow CH_4 \quad \Delta H_r = -74.7 \ \text{MJ/kmol} \tag{6}$$

The designation C represents carbon in the original feedstock as well as carbon in the char formed by devolatilization of the feedstock. Reactions (1) and (2) are exothermic oxidation reactions and provide most of the energy required by the endothermic gasification reactions (3) and (4). The oxidation reactions occur very rapidly, completely consuming all of the oxygen present in the gasifier, so that most of the gasifier operates under reducing conditions. Reaction (5) is the water − gas shift reaction, in water (steam) is converted to hydrogen − this reaction is used to alter the hydrogen/carbon monoxide ration when synthesis gas is the desired product, such as for use in Fischer − Tropsch processes. Reaction (6) is favored by high pressure and low temperature and is, thus, mainly important in lower temperature gasification systems. Methane formation is an exothermic reaction that does not consume oxygen and, therefore, increases the efficiency of the gasification process and the final heat content of the product gas. Overall, approximately 70% of the heating value of the product gas is associated with the carbon monoxide and hydrogen but this varies depending on the gasifier type and the process parameters (Speight, 2011a; Chadeesingh, 2011; Speight, 2013a).

In essence, the direction of the gasification process is subject to the constraints of thermodynamic equilibrium and variable reaction kinetics. The combustion reactions (reaction of the feedstock or char with oxygen) essentially go to completion. The thermodynamic equilibrium of the rest of the gasification reactions are relatively well defined and collectively have a major influence on thermal efficiency of the process as well as on the gas composition. Thus, thermodynamic data are useful for estimating key design parameters for a gasification process, such as: (1) calculating the relative amounts of oxygen and/or steam required per unit of feedstock, (2) estimating the composition of the produced synthesis gas, and (3) optimizing process efficiency at various operating conditions.

Other deductions concerning gasification process design and operations can also be derived from the thermodynamic understanding of its reactions. Examples include: (1) production of synthesis gas with low methane content at high temperature, which requires an amount of steam in excess of the stoichiometric requirement, (2) gasification at high temperature, which increases oxygen consumption and decreases the overall process efficiency, and (3) production of synthesis gas with a high methane content, which requires operation at low temperature (approximately 700°C, 1290°F) but the methanation reaction kinetics will be poor without the presence of a catalyst.

Relative to the thermodynamic understanding of the gasification process, the kinetic behavior is much more complex. In fact, very little reliable global kinetic information on gasification reactions exists, partly because it is highly dependent on: (1) the chemical nature of the feed, which varies significantly with respect to composition, mineral impurities, (2) feedstock reactivity, and (3) process conditions. In addition, physical characteristics of the feedstock (or char) also play a role in phenomena such as boundary layer diffusion, pore diffusion, and ash layer diffusion, which also influence the kinetic outcome. Furthermore, certain impurities are known to have catalytic activity on some of the gasification reactions, which can have a further influence on the kinetic imprint of the gasification reactions.

2.3.1 Primary Gasification

Primary gasification involves thermal decomposition of the raw feedstock via various chemical processes and many schemes involve pressures ranging from atmospheric to 1000 psi. Air or oxygen may be admitted to support combustion to provide the necessary heat. The product is usually a low-heat-content (low-Btu) gas ranging from a carbon monoxide/hydrogen mixture to mixtures containing varying amounts of carbon monoxide, carbon dioxide, hydrogen, water, methane, hydrogen sulfide, and nitrogen, and typical tar-like products of thermal decomposition of carbonaceous feedstocks are complex mixtures and include hydrocarbon oils and phenolic products (Dutcher et al., 1983; Speight, 2011a, 2013a, 2013b).

Devolatilization of the feedstock occurs rapidly as the temperature rises above 300°C (570°F). During this period, the chemical structure is altered, producing solid char, tar products, condensable liquids, and low molecular weight gases. Furthermore, the products of the devolatilization stage in an inert gas atmosphere are very different from those in an atmosphere containing hydrogen at elevated pressure. In an atmosphere of hydrogen at elevated pressure, additional yields of methane or other low molecular weight gaseous hydrocarbons can result during the initial gasification stage from reactions such as: (1) direct hydrogenation of feedstock or semi-char because of any reactive intermediates formed and (2) the hydrogenation of other gaseous hydrocarbons, oils, tars, and carbon oxides. Again, the kinetic picture for such reactions is complex due to the varying composition of the volatile products which, in turn, are related to the chemical character of the feedstock and the process parameters, including the reactor type.

A solid char product may also be produced, and may represent the bulk of the weight of the original feedstock, which determines (to a large extent) the yield of char and the composition of the gaseous product.

2.3.2 Secondary Gasification

Secondary gasification usually involves the gasification of char from the primary gasifier, which is typically achieved by reaction of the hot char with water vapor to produce carbon monoxide and hydrogen:

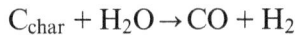

$$C_{char} + H_2O \rightarrow CO + H_2$$

The reaction requires heat input (endothermic) for the reaction to proceed in its forward direction. Usually, an excess amount of steam is also needed to promote the reaction. However, excess steam used in this reaction has an adverse effect on the thermal efficiency of the process. Therefore, this reaction is typically combined with other gasification reactions in practical applications. The hydrogen − carbon monoxide ratio of the product synthesis gas depends on the synthesis chemistry as well as process engineering.

The mechanism of this reaction is based on the reaction between carbon and gaseous reactants, not for reactions between feedstock and gaseous reactants. Hence the equations may over-simply the actual chemistry of the steam gasification reaction. Even though carbon is the dominant atomic species present in feedstock, feedstock is more reactive than pure carbon. The presence of various reactive organic functional groups and the availability of catalytic activity via naturally occurring mineral ingredients can enhance the relative reactivity of the feedstock − for example anthracite, which has the highest carbon content among all ranks of coal (Speight, 2013a), is most difficult to gasify or liquefy.

Once the rate of devolatilization has passed a maximum another reaction becomes important − in this reaction the semi-char is converted to char (sometimes erroneously referred to as *stable char*) primarily through the evolution of hydrogen. Thus, the gasification process occurs as the char reacts with gases such as carbon dioxide and steam to produce carbon monoxide and hydrogen. The resulting gas (producer gas or synthesis gas) may be more efficiently converted to electricity than is typically possible by direct combustion of the coal. Also, corrosive elements in the ash may be refined out by the gasification process, allowing high temperature combustion of the gas from otherwise problematic feedstocks (Speight, 2011a, 2013a, 2013b).

Oxidation and gasification reactions consume the char and the oxidation and the gasification kinetic rates follow Arrhenius type dependence on temperature. On the other hand, the kinetic parameters are feedstock specific and there is no true global relationship to describe the kinetics of char gasification – the characteristics of the char are also feedstock specific. The complexity of the reactions makes the reaction initiation and the subsequent rates subject to many factors, any one of which can influence the kinetic aspects of the reaction.

Although the initial gasification stage (devolatilization) is completed in seconds or even less at elevated temperature, the subsequent gasification of the char produced at the initial gasification stage is much slower, requiring minutes or hours to obtain significant conversion under practical conditions and reactor designs for commercial gasifiers are largely dependent on the reactivity of the char and also on the gasification medium (Johnson, 1979; Penner, 1987; Sha, 2005). Thus, the distribution and chemical composition of the products are also influenced by the prevailing conditions (i.e., temperature, heating rate, pressure, residence time, etc.) and, last but not least, the nature of the feedstock. Also, the presence of oxygen, hydrogen, water vapor, carbon oxides, and other compounds in the reaction atmosphere during pyrolysis may either support or inhibit numerous reactions with feedstock and with the products evolved.

The reactivity of char produced in the pyrolysis step depends on nature of the feedstock and increases with oxygen content of the feedstock but decreases with carbon content. In general, char produced from a low-carbon feedstock is more reactive than char produced from a high-carbon feedstock. The reactivity of char from a low-carbon feedstock may be influenced by catalytic effect of mineral matter in char. In addition, as the carbon content of the feedstock increases, the reactive functional groups present in the feedstock decrease and the char becomes more aromatic and cross-linked in nature (Speight, 2013a). Therefore char obtained from high-carbon feedstock contains a lesser number of functional groups and higher proportion of aromatic and cross-linked structures, which reduce reactivity. The reactivity of char also depends upon thermal treatment it receives during formation from the parent feedstock – the gasification rate of char decreases as the char preparation temperature increases due to the decrease in active surface areas of char. Therefore a change of char preparation temperature may change the chemical nature of char, which in turn may change the gasification rate.

Typically, char has a higher surface area compared to the surface area of the parent feedstock, even when the feedstock has been pelletized, and the surface area changes as the char undergoes gasification – the surface area increases with carbon conversion, reaches a maximum, and then decreases. These changes in turn affect gasification rates – in general, reactivity increases with the increase in surface area. The initial increase in surface area appears to be caused by cleanup and widening of pores in the char. The decrease in surface area at high carbon conversion may be due to coalescence of pores, which ultimately leads to collapse of the pore structure within the char.

Heat transfer and mass transfer processes in fixed or moving bed gasifiers are affected by complex solids flow and chemical reactions. Coarsely crushed feedstock settles while undergoing heating, drying, devolatilization, gasification, and combustion. Also, the feedstock particles change in diameter, shape, and porosity – non-ideal behavior may result from certain types of chemical structures in the feedstock, gas bubbles, and channel and a variable void fraction may also change heat and mass transfer characteristics.

Pyrolysis temperature is an important factor in the thermal history, and consequently in the thermodynamics of the feedstock char. However, the thermal history of a char should also depend on the rate of temperature rise to the pyrolysis temperature and on the length of time the char is kept at the pyrolysis temperature (soak time), which might be expected to reduce the residual entropy of the char by employing a longer soak time.

Alkali metal salts are known to catalyze the steam gasification reaction of carbonaceous materials, including coal. The process is based on the concept that alkali metal salts (such as potassium carbonate, sodium carbonate, potassium sulfide, sodium sulfide, and the like) will catalyze the steam gasification of feedstocks. The order of catalytic activity of alkali metals on the gasification reaction is:

Cesium(Cs) > rubidium(Rb) > potassium(K) > sodium(Na) > lithium(Li)

Catalyst amounts on the order of 10 to 20% w/w potassium carbonate will lower bituminous coal gasifier temperatures from 925°C (1695°F) to 700°C (1090°F) and means that the catalyst can be introduced to the gasifier impregnated on coal or char. In addition, tests with potassium carbonate showed that this material also acts as a catalyst for the methanation reaction. Furthermore, the use of catalysts can reduce the amount of tar formed in the process (Cusumano et al., 1978; McKee, 1981; Shinnar et al., 1982). In the case of catalytic steam gasification of coal, carbon deposition reaction may affect catalyst life by fouling the catalyst active sites. This carbon deposition reaction is more likely to take place whenever the steam concentration is low.

Ruthenium-containing catalysts are used primarily in the production of ammonia. It has been shown that ruthenium catalysts provide five to 10 times higher reactivity rates than other catalysts. However, ruthenium quickly becomes inactive due to its necessary supporting material, such as activated carbon, which is used to achieve effective reactivity. However, during the process, the carbon is consumed, thereby reducing the effect of the ruthenium catalyst.

Catalysts can also be used to favor or suppress the formation of certain components in the gaseous product by changing the chemistry of the reaction, the rate of reaction, and the thermodynamic balance of the reaction. For example, in the production of synthesis gas (mixtures of hydrogen and carbon monoxide), methane is also produced in small amounts. Catalytic gasification can be used to either promote methane formation or suppress it.

2.3.3 Water − Gas Shift Reaction

The water − gas shift reaction (shift conversion) is necessary because the gaseous product from a gasifier generally contains large amounts of carbon monoxide and hydrogen, plus lesser amounts of other gases. Carbon monoxide and hydrogen (if they are present in the mole ratio of 1:3) can be reacted in the presence of a catalyst to produce methane. However, some adjustment to the ideal (1:3) is usually required and, to accomplish this, all or part of the stream is treated according to the waste gas shift (shift conversion) reaction. This involves reacting carbon monoxide with steam to produce a carbon dioxide and hydrogen whereby the desired 1:3 mole ratio of carbon monoxide to hydrogen may be obtained:

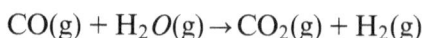

$$CO(g) + H_2O(g) \rightarrow CO_2(g) + H_2(g)$$

Even though the water − gas shift reaction is not classified as one of the principal gasification reactions, it cannot be omitted in the analysis of chemical reaction systems that involve synthesis gas. Among all reactions involving synthesis gas, this reaction equilibrium is least sensitive to the temperature variation − the equilibrium constant is least strongly dependent on the temperature. Therefore, the reaction equilibrium can be reversed in a variety of practical process conditions over a wide range of temperatures.

The water − gas shift reaction in its forward direction is mildly exothermic and although all of the participating chemical species are in gaseous form, the reaction is believed to be heterogeneous insofar as the chemistry occurs at the surface of the feedstock and the reaction is actually catalyzed by carbon surfaces. In addition, the reaction can also take place homogeneously as well as heterogeneously and a generalized understanding of the water gas shift reaction is difficult to achieve. Even the published kinetic rate information is not immediately useful or applicable to a practical reactor situation.

Synthesis gas from a gasifier contains a variety of gaseous species other than carbon monoxide and hydrogen. Typically, they include carbon dioxide, methane, and water (steam) (Cusumano et al., 1978). Depending on the objective of the ensuing process, the composition of synthesis gas may need to be preferentially readjusted. If the objective of the gasification process is to obtain a high yield of methane, it would be preferred to have the molar ratio of hydrogen to carbon monoxide at 3:1:

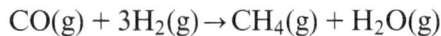

$$CO(g) + 3H_2(g) \rightarrow CH_4(g) + H_2O(g)$$

On the other hand, if the objective of generating synthesis gas is the synthesis of methanol via a vapor-phase low-pressure process, the stoichiometrically consistent ratio between hydrogen and carbon monoxide would be 2:1. In such cases, the stoichiometrically consistent synthesis gas mixture is often referred to as *balanced gas*, whereas a synthesis gas composition that is substantially deviated from the principal reaction's stoichiometry is called *unbalanced gas*. If the objective of synthesis gas production is to obtain a high yield of hydrogen, it would be advantageous to increase the ratio of hydrogen to carbon monoxide by further converting carbon monoxide (and water) into hydrogen (and carbon dioxide) via the water − gas shift reaction.

The water – gas shift reaction is one of the major reactions in the steam gasification process, where both water and carbon monoxide are present in ample amounts. Although the four chemical species involved in the water – gas shift reaction are gaseous compounds at the reaction stage of most gas processing, the water – gas shift reaction, in the case of steam gasification of feedstock, predominantly takes place on the solid surface of feedstock (heterogeneous reaction). If the product synthesis gas from a gasifier needs to be reconditioned by the water – gas shift reaction, this reaction can be catalyzed by a variety of metallic catalysts.

Choice of specific kinds of catalysts has always depended on the desired outcome, the prevailing temperature conditions, composition of gas mixture, and process economics. Typical catalysts used for the reaction include those containing iron, copper, zinc, nickel, chromium, and molybdenum.

2.3.4 Carbon Dioxide Gasification

The reaction of carbonaceous feedstocks with carbon dioxide produces carbon monoxide (*Boudouard reaction*) and (like the steam gasification reaction) is also an endothermic reaction:

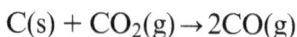

$$C(s) + CO_2(g) \rightarrow 2CO(g)$$

The reverse reaction results in carbon deposition (carbon fouling) on many surfaces including the catalysts and results in catalyst deactivation.

This gasification reaction is thermodynamically favored at high temperatures ($>680°C$, $>1255°F$), which is also quite similar to steam gasification. If carried out alone, the reaction requires high temperature (for fast reaction) and high pressure (for higher reactant concentrations) for significant conversion but as a separate reaction a variety of factors come into play: (1) low conversion, (2) slow kinetic rate, and (3) low thermal efficiency.

Also, the rate of the carbon dioxide gasification of a feedstock is different to the rate of the carbon dioxide gasification of carbon. Generally, the carbon – carbon dioxide reaction follows a reaction order based on the partial pressure of the carbon dioxide that is approximately 1.0 (or lower) whereas the feedstock – carbon dioxide reaction follows a reaction order based on the partial pressure of the carbon dioxide that is 1.0 (or higher). The observed higher reaction order for the feedstock reaction is also based on the relative reactivity of the feedstock in the gasification system.

2.3.5 Hydrogasification

Not all high-heat-content (high-Btu) gasification technologies depend entirely on catalytic methanation and, in fact, a number of gasification processes use hydrogasification, that is, the direct addition of hydrogen to feedstock under pressure to form methane (Anthony and Howard, 1976).

$$C_{char} + 2H_2 \rightarrow CH_4$$

The hydrogen-rich gas for hydrogasification can be manufactured from steam and char from the hydrogasifier. Appreciable quantities of methane are formed directly in the primary gasifier and the heat released by methane formation is at a sufficiently high temperature to be used in the steam – carbon reaction to produce hydrogen so that less oxygen is used to produce heat for the steam – carbon reaction. Hence, less heat is lost in the low-temperature methanation step, thereby leading to higher overall process efficiency.

Hydrogasification is the gasification of feedstock in the presence of an atmosphere of hydrogen under pressure (Anthony and Howard, 1976). Thus, not all high-heat-content (high-Btu) gasification technologies depend entirely on catalytic methanation and, in fact, a number of gasification processes use hydrogasification, that is, the direct addition of hydrogen to feedstock under pressure to form methane:

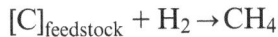

$$[C]_{feedstock} + H_2 \rightarrow CH_4$$

The hydrogen-rich gas for hydrogasification can be manufactured from steam by using the char that leaves the hydrogasifier. Appreciable quantities of methane are formed directly in the primary gasifier and the heat released by methane formation is at a sufficiently high temperature to be used in the steam – carbon reaction to produce hydrogen so that less oxygen is used to produce heat for this reaction. Hence, less heat is lost in the low-temperature methanation step, thereby leading to higher overall process efficiency.

The hydrogasification reaction is exothermic and is thermodynamically favored at low temperatures ($<670°C$, $<1240°F$), unlike the endothermic steam gasification and carbon dioxide gasification reactions. However, at low temperatures, the reaction rate is inevitably too slow. Therefore, a high temperature is always required for kinetic reasons, which in turn requires high pressure of hydrogen, which is also preferred from equilibrium considerations. This reaction can be catalyzed by salts such as potassium carbonate (K_2CO_3), nickel chloride ($NiCl_2$), iron chloride ($FeCl_2$), and iron sulfate ($FeSO_4$). However, use of a catalyst in feedstock gasification suffers from difficulty in recovering and reusing the catalyst and the potential for the spent catalyst to become an environmental issue.

In a hydrogen atmosphere at elevated pressure, additional yields of methane or other low molecular weight hydrocarbons can result during the initial feedstock gasification stage from direct hydrogenation of feedstock or semi-char because of active intermediate formed in the feedstock structure after pyrolysis. The direct hydrogenation can also increase the amount of feedstock carbon that is gasified as well as the hydrogenation of gaseous hydrocarbons, oil, and tar.

The kinetics of the rapid-rate reaction between gaseous hydrogen and the active intermediate depends on hydrogen partial pressure (P_{H2}). Greatly increased gaseous hydrocarbons produced during the initial feedstock gasification stage are extremely important in processes to convert feedstock into methane (SNG, synthetic natural gas).

2.3.6 Methanation

Several exothermic reactions may occur simultaneously within a methanation unit. A variety of metals have been used as catalysts for the methanation reaction; the most common, and to some extent the most effective methanation catalysts, appear to be nickel and ruthenium, with nickel being the most widely used (Cusumano et al., 1978):

Ruthenium(Ru) > nickel(Ni) > cobalt(Co) > iron(Fe) > molybdenum(Mo).

Nearly all the commercially available catalysts used for this process are, however, very susceptible to sulfur poisoning and efforts must be taken to remove all hydrogen sulfide (H_2S) before the catalytic reaction starts. It is necessary to reduce the sulfur concentration in the feed gas to less than 0.5 ppm v/v in order to maintain adequate catalyst activity for a long period of time.

The synthesis gas must be desulfurized before the methanation step since sulfur compounds will rapidly deactivate (poison) the catalysts.

A problem may arise when the concentration of carbon monoxide is excessive in the stream to be methanated since large amounts of heat must be removed from the system to prevent high temperatures and deactivation of the catalyst by sintering as well as the deposition of carbon. To eliminate this problem, temperatures should be maintained below 400°C (750°F).

The methanation reaction is used to increase the methane content of the product gas, as needed for the production of high-Btu gas:

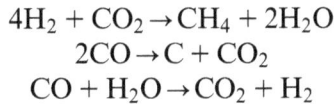

$$4H_2 + CO_2 \rightarrow CH_4 + 2H_2O$$
$$2CO \rightarrow C + CO_2$$
$$CO + H_2O \rightarrow CO_2 + H_2$$

Among these, the most dominant chemical reaction leading to methane is the first one. Therefore, if methanation is carried out over a catalyst with a synthesis gas mixture of hydrogen and carbon monoxide, the desired hydrogen − carbon monoxide ratio of the feed synthesis gas is around 3:1. The large amount of water (vapor) produced is removed by condensation and recirculated as process water or steam. During this process, most of the exothermic heat due to the methanation reaction is also recovered through a variety of energy integration processes.

Whereas all the reactions listed above are quite strongly exothermic except the forward water − gas shift reaction, which is mildly exothermic, the heat release depends largely on the amount of carbon monoxide present in the feed synthesis gas. For each 1% v/v carbon monoxide in the feed synthesis gas, an adiabatic reaction will experience a 60°C (108°F) temperature rise, which may be termed as *adiabatic temperature rise*.

3 PRODUCTS

If air is used for combustion, the product gas will have a heat content on the order of 150 to 300 Btu/ft3 depending on process design characteristics and will contain undesirable constituents such as carbon dioxide, hydrogen sulfide, and nitrogen. The use of pure oxygen results in a product gas having a heat content of 300 to 400 Btu/ft3 with carbon dioxide and hydrogen sulfide as byproducts, both of which can be removed from low-heat-content or medium-heat-content − low-Btu or

Table 2.2 Gasification Products	
Product	Characteristics
Low-Btu gas (150–300 Btu/scf)	Around 50% nitrogen, with smaller quantities of combustible H_2 and CO, CO_2 and trace gases, such as methane
Medium-Btu gas (300–550 Btu/scf)	Predominantly CO and H_2, with some incombustible gases and sometimes methane
High-Btu gas (980–1080 Btu/scf)	Almost pure methane

medium-Btu gas (Table 2.2) – by any of several available processes (see Table 2.1) (Mokhatab et al., 2006; Speight, 2013a, 2014).

If high-heat-content (high-Btu) gas (900 to 1000 Btu/ft^3) is required, efforts must be made to increase the methane content of the gas. The reactions that generate methane are all exothermic and have negative values, but the reaction rates are relatively slow and catalysts may, therefore, be necessary to accelerate the reactions to acceptable commercial rates. Indeed, it is also possible that the mineral constituents of the feedstock and char may modify the reactivity by a direct catalytic mechanism. The presence of oxygen, hydrogen, water vapor, carbon oxides, and other compounds in the reaction atmosphere during pyrolysis may either support or inhibit numerous reactions with the feedstock and with the products evolved.

If high-Btu gas (high-heat-content gas; 900 to 1000 Btu/ft^3) is the desired product, efforts must be made to increase the methane content of the gas. The reactions that generate methane are all exothermic and have negative values (Lee, 2007), but the reaction rates are relatively slow and catalysts may, therefore, be necessary to accelerate the reactions to acceptable commercial rates.

3.1 Effect of Process Parameters

Depending on the type of feedstock being gasified and the analysis of the gas product desired, pressure also plays a role in product definition. In fact, prior to gasification, some (or all) of the following processing steps will be required: (1) pretreatment of the feedstock, (2) primary gasification, (3) secondary gasification of the carbonaceous residue from the primary gasifier, (4) removal of carbon dioxide, hydrogen sulfide, and other acid gases, (5) shift conversion for adjustment of the carbon monoxide/hydrogen mole ratio to the desired ratio, and (6) catalytic methanation of the carbon monoxide/hydrogen mixture to form methane. If high-heat-content (high-Btu) gas is desired, all of these processing steps are required since gasifiers (irrespective of the feedstock) do not typically yield methane in the concentrations required (Mills, 1969; Graff et al., 1976; Cusumano et al., 1978; Mills, 1982).

3.2 Thermodynamics and Kinetics

Relative to the chemical and thermodynamic understanding of the gasification process and data derived from thermodynamic studies (van der Burgt, 2008; Shabbar and Janajreh, 2013), the kinetic behavior of carbonaceous or hydrocarbonaceous feedstocks is more complex.

The chemistry of gasification is quite complex and, only for discussion purposes, can the chemistry be viewed as consisting of a few major reactions which can progress to different extents depending on the gasification conditions (such as temperature and pressure) and the feedstock used. Combustion reactions take place in a gasification process, but, in comparison with conventional combustion which uses a stoichiometric excess of oxidant, gasification typically uses one-fifth to one-third of the theoretical oxidant. This only partially oxidizes the carbon feedstock. As a *partial oxidation* process, the major combustible products of gasification are carbon monoxide (CO) and hydrogen, with only a minor portion of the carbon completely oxidized to carbon dioxide (CO_2). The heat produced by the partial oxidation provides most of the energy required to drive the endothermic gasification reactions.

Furthermore, while the basic thermodynamic cycles pertinent to the gasification process have long been established, novel combination and the use of alternative fluids to water/steam offer the prospect of higher process efficiency through use of thermodynamic studies.

Finally, very little reliable kinetic information on gasification reactions exists, partly because it is highly dependent on the process conditions and the nature of the feedstock, which can vary significantly with respect to composition, mineral impurities, and reactivity as well as the potential for certain impurities to exhibit catalytic activity on some of the gasification reactions. Indeed, in spite of the efforts of many researchers, they are far from being able to apply kinetic data to gasification of a carbonaceous feedstock or char in various processes.

Gasification is one of the critical technologies that enable hydrogen production from solid hydrocarbons such as coal and biomass (Speight, 2008). Gasifiers produce a synthesis gas that has multiple applications and can be used for hydrogen production, electricity generation, and chemical plants. Integrated gasification combined cycle (IGCC) plants utilize the synthesis gas in a combined cycle power plant (gas turbine and steam turbine) to produce electricity (Speight, 2013b).

With the increasing costs of petroleum, the gasification-based refinery is another concept for the production of fuels, electricity, and chemical products (Speight, 2011b). Coal gasification has also been used for production of liquid fuels (Fischer − Tropsch diesel and methanol) via a catalytic conversion of synthesis gas into liquid hydrocarbons (Speight, 2008, 2011a; Chadeesingh, 2011).

4 CATALYTIC GASIFICATION

Catalysts are commonly used in the chemical and petroleum industries to increase reaction rates, sometimes making certain previously unachievable products possible (Speight and Ozum, 2002; Hsu and Robinson, 2006; Speight, 2014). Acids, through donated protons (H^+), are common reaction catalysts, especially in the organic chemical industries. Thus it is not surprising that catalysts can be used to enhance the reactions involved in gasification and use of appropriate catalysts not only reduces reaction temperature but also improves the gasification rates.

In addition, thermodynamic constraints of the gasification process that limit the thermal efficiency are not inherent but the result of design decisions based on available technology, as well as the kinetic properties of available catalysts. The availability of relevant catalysts limits the yield of methane to that obtainable at global equilibrium over carbon in the presence of carbon monoxide and hydrogen. The equilibrium composition is shown to be independent of the thermodynamic properties of the char or feedstock. These limitations give nonisothermal two-stage processes significant thermodynamic advantages. The results of analysis suggest directions for modifying present processes to obtain higher thermal efficiencies, and two-stage process schemes that would have significant advantages over present technologies and should be applicable to a wide range of catalytic and non-catalytic processes (Shinnar et al., 1982; McKee, 1981).

Alkali metal salts of weak acids, such as potassium carbonate (K_2CO_3), sodium carbonate (Na_2CO_3), potassium sulfide (K_2S), and sodium sulfide (Na_2S) can catalyze the carbon – steam gasification reaction.

Ruthenium-containing catalysts are used primarily in the production of ammonia. It has been shown that ruthenium catalysts provide five to 10 times higher reactivity rates than other catalysts. However, ruthenium quickly becomes inactive due to the need for a supporting material, such as activated carbon, which is used to achieve effective reactivity. During the process, the carbon is consumed, thereby reducing the effect of the ruthenium catalyst.

Catalysts can also be used to favor or suppress the formation of certain components in the gaseous product. For example, in the production of synthesis gas (mixtures of hydrogen and carbon monoxide), methane is also produced in small amounts. Catalytic gasification can be used to either promote methane formation or suppress it.

Disadvantages of catalytic gasification include increased material costs for the catalyst itself (often rare metals), as well as diminishing catalyst performance over time. Catalysts can be recycled, but their performance tends to diminish with age or by poisoning. The relative difficulty in reclaiming and recycling the catalyst can also be a disadvantage. For example, the potassium carbonate catalyst can be recovered from spent char with a simple water wash, but other catalysts may not be so accommodating. Many catalysts are sensitive to particular chemical species which bond with the catalyst or alter it in such a way that it no longer functions. Sulfur, for example, can poison several types of catalysts including palladium and platinum.

REFERENCES

Anthony, D.B., Howard, J.B., 1976. Coal devolatilization and hydrogasification. Am. Inst. Chem. Eng. J. 22 (4), 625–656.

Baker, R.T.K., Rodriguez, N.M., 1990. In Fuel Science and Technology Handbook. Marcel Dekker Inc, New York (Chapter 22).

Chadeesingh, R., 2011. The fischer – tropsch process. In: Speight, J.G. (Ed.), The Biofuels Handbook. The Royal Society of Chemistry, London, UK, pp. 476–517. (Part 3, Chapter 5).

Cusumano, J.A., Dalla Betta, R.A., Levy, R.B., 1978. Catalysis in Coal Conversion. Academic Press Inc., New York.

Davidson, R.M., 1983. Mineral Effects in Coal Conversion. Report No. ICTIS/TR22, International Energy Agency, London, UK.

Dutcher, J.S., Royer, R.E., Mitchell, C.E., Dahl, A.R., 1983. In: Wright, C.W., Weimer, W.C., Felic, W.D. (Eds.), Advanced Techniques in Synthetic Fuels Analysis. Technical Information Center, United States Department of Energy, Washington, DC, p. 12.

Graff, R.A., Dobner, S., Squitres, A.M., 1976. Flash Hydrogenation of Coal. 1. Experimental Methods and Results. 2. Yield Structure for Illinois No. 6 Coal at 100 Atm. Fuel, 55, 109−113.

Higman, C., Van der Burgt, M., 2008. Gasification 2nd Edition. Gulf Professional Publishing, Elsevier, Amsterdam, Netherlands.

Hsu, C.S., Robinson, P.R., 2006. Practical Advances in Petroleum Processing, Volumes 1 and 2. Springer, New York.

Irfan, M.F., 2009. Research Report: Pulverized Coal Pyrolysis & Gasification in $N_2/O_2/CO_2$ Mixtures by Thermo-gravimetric Analysis, Volume 2. Novel Carbon Resource Sciences Newsletter, Kyushu University, Fukuoka, Japan, 27−33.

Johnson, J.L., 1979. Kinetics of Coal Gasification. John Wiley and Sons Inc, New York.

Lee, S., 2007. Gasification of coal. In: Lee, S., Speight, J.G., Loyalka, S. (Eds.), Handbook of Alternative Fuel Technologies, 2007. CRC Press, Taylor & Francis Group, Boca Raton, FL, pp. 25−80.

Martinez-Alonso, A., Tascon, J.M.D., 1991. Determining the role of mineral matter on gasification reactivity of brown coal chars. In: Lahaye, J., Ehrburger, P. (Eds.), Fundamental Issues in Control of Carbon Gasification Reactivity. Kluwer Academic Publishers, Dordrecht, Netherlands.

McKee, D.W., 1981. Mechanisms of Catalyzed Gasification of Carbon. Proceedings. AIP Conference, American Institute of Physics, College Park, Maryland. Volume 70, p. 236−255.

Mills, G.A., 1969. Ind. Eng. Chem. 61 (7), 6.

Mills, G.A., 1982. Chemtech 12, 294.

Mims, C.A., 1991. Catalytic gasification of coal: fundamentals and mechanism. In: Lahaye, J., Ehrburger, P. (Eds.), Fundamental Issues in Control of Carbon Gasification Reactivity. Kluwer Academic Publishers, Dordrecht, Netherlands, p. 383.

Mokhatab, S., Poe, W.A., Speight, J.G., 2006. Handbook of Natural Gas Transmission and Processing. Elsevier, Amsterdam, Netherlands.

Müller, R., von Zedtwitz, P., Wokaun, A., Steinfeld, A., 2003. Kinetic investigation on steam gasification of charcoal under high-flux radiation. Chem. Eng. Sci. 58, 5111−5119.

Penner, S.S., 1987. Coal Gasification. Pergamon Press Inc, New York.

Pepiot, P., Dibble, C.J., Foust, C.G., 2010. Computational fluid dynamics modeling of biomass gasification and pyrolysis. In: Nimlos, M.R., Crowley, M.F. (Eds.), Computational Modeling in Lignocellulosic Biofuel Production. ACS Symposium Series; American Chemical Society, Washington, DC.

Sha, X., 2005. Coal gasification. In: Coal, Oil Shale, Natural Bitumen, Heavy Oil and Peat. Encyclopedia of Life Support Systems (EOLSS), Developed under the Auspices of the UNESCO, EOLSS Publishers, Oxford, UK <http://www.eolss.net>.

Shabbar, S., Janajreh, I., 2013. Thermodynamic equilibrium analysis of coal gasification using gibbs energy minimization method. Energy Conversion Manage. 65, 755−763.

Shinnar, R., Fortuna, G., Shapira, D., 1982. Thermodynamic and Kinetic Constraints of Catalytic Synthetic Natural gas Processes. 21: 728−750.

Singh, S.P., Weil, S.A., Babu, S.P., 1980. Thermodynamic analysis of coal gasification processes. Energy 5 (8-9), 905–914.

Slavinskaya, N.A., Petrea, D.M., Riedel. U. 2009. Chemical Kinetic Modeling in Coal Gasification Overview. Proceedings 5th International Workshop on Plasma Assisted Combustion (IWEPAC). Alexandria, Virginia.

Speight, J.G., 2008. Synthetic Fuels Handbook: Properties, Processes, and Performace. McGraw-Hill, New York.

Speight, J.G. (Ed.), 2011a. The Biofuels Handbook. The Royal Society of Chemistry, London, UK.

Speight, J.G., 2011b. The Refinery of the Future. Gulf Professional Publishing, Elsevier, Oxford, UK.

Speight, J.G., 2013a. The Chemistry and Technology of Coal 3rd Edition. CRC Press, Taylor & Francis Group, Boca Raton, FL.

Speight, J.G., 2013b. Coal-Fired Power Generation Handbook. Scrivener Publishing, Salem, MA.

Speight, J.G., 2014. The Chemistry and Technology of Petroleum 5th Edition. CRC Press, Taylor & Francis Group, Boca Raton, FL.

Speight, J.G., Ozum, B., 2002. Petroleum Refining Processes. Marcel Dekker Inc., New York.

Sundaresan, S., Amundson, N.R., 1978. Studies in char gasification – I: a lumped model. Chem. Eng. Sci. 34, 345–354.

Van der Burgt, M., 2008. The thermodynamics of gasification. In: Higman, C., van der Burgt, M. (Eds.), Gasification 2nd Edition. Gulf Professional Publishing, Elsevier, Amsterdam, Netherlands, Chapter 2.

Wang, W., Mark, T.K., 1992. Fuel. 71: 871.

CHAPTER 3

Gasifier Types

1 INTRODUCTION

There has been a general tendency to classify gasification processes by virtue of the heat content of the gas which is produced but it is also possible, and often more appropriate, to classify gasification processes according to the type of reactor vessel and whether or not the system reacts under pressure. However, for the purposes of the present text gasification processes are segregated according to the bed types, which differ in their ability to accept (and convert) various types of feedstock (Figure 3.1) (Collot, 2002, 2006). The main differences in the gasifiers are:

1. The method by which the feedstock is introduced into the gasifier and is moved around within it − the feedstock is either fed into the top of the gasifier, or into the side, and then is moved around either by gravity or air flows.
2. The use of an oxidant such as oxygen, air or steam− using air dilutes the synthesis gas with nitrogen, which adds to the cost of downstream processing while using oxygen avoids this, but is expensive, and so oxygen-enriched air can also be used.
3. The temperature range in which the gasifier is operated.
4. The means by which the heat for the gasifier is provided by partially combusting some of the biomass in the gasifier (directly heated), or from an external source (indirectly heated), such as circulation of an inert material or steam.
5. The pressure at which the gasifier is operated − above atmospheric pressure, which provides a higher throughput with larger maximum capacities, promotes hydrogen production and leads to smaller, cheaper downstream cleanup equipment; since no additional compression is required, the synthesis gas temperature can be kept high for downstream operations and liquid fuels catalysis but at pressures in excess of 450 psi costs quickly increase, since gasifiers need to be more robustly engineered, and the required feeding mechanisms involve complex pressurizing steps.

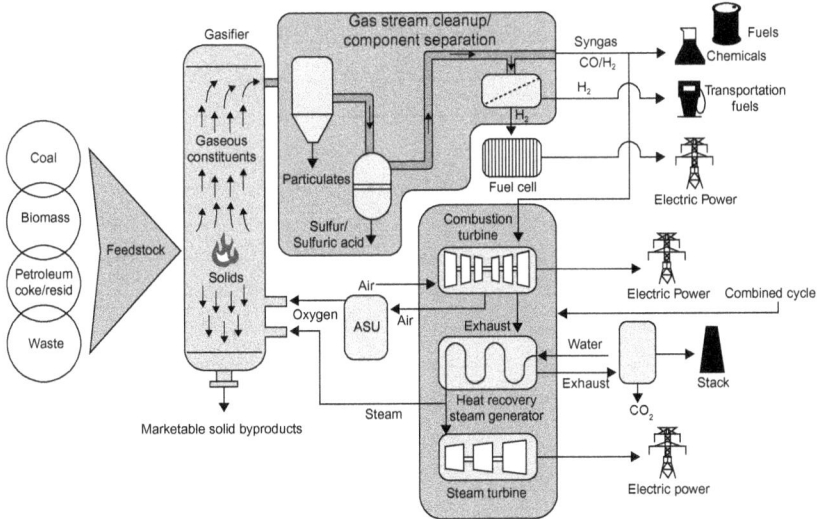

Figure 3.1 Gasification flow for various feedstocks. ASU, air separation unit.

Although there are many successful commercial gasifiers, the basic form and concept are available but details on the design and operation for the commercial coal gasifiers are closely guarded as proprietary information. In fact, the production of gas from carbonaceous feedstocks has been an expanding area of technology. As a result, several types of gasification reactors have arisen and there has been a general tendency to classify gasification processes by virtue of the heat content of the gas which is produced (Collot, 2002, 2006).

It is the purpose of this chapter to present the different categories of gasification reactors as they apply to various types of feedstocks. Within each category there are several commonly known processes – represented through coal gasification processes (Table 3.1) – some of which are in current use and some of which are in lesser use (Speight, 2008, 2011, 2013a, 2013b).

2 GASIFIER DESIGN

Several types of fuels are available for gasification and include coal, petroleum residua, wood and wood waste (branches, twigs, roots, bark, wood shavings, and sawdust) as well as a multitude of agricultural residues (maize cobs, coconut shells, coconut husks, cereal straws, rice husks, etc.) and peat (Chapter 1). Because the fuels differ greatly in their

Table 3.1 Categories of Coal Gasification Processes

Fixed bed processes
Foster Wheeler stoic process
Lurgi process
Wellman Galusha process
Woodall – Duckham process
Fluidized bed processes
Agglomerating burner process
Carbon dioxide acceptor process
Coalcon process
COED/COGAS process
Exxon catalytic gasification process
Hydrane process
Hygas process
Pressurized fluid bed process
Synthane process
U-gas process
Winkler process
Entrained bed processes
Bi-gas process
Combustion engineering process
Koppers – Totzek process
Texaco process
Molten salt processes
Atgas process
Pullman – Kellogg process
Rockgas process
Rummel single-shaft process
From Braunstein et al. (1977).

chemical, physical, and morphological properties, they make different demands on the method of gasification and consequently require different reactor design and/or on the gasification technology. It is for this reason that, during more than a century of gasification experience, a large number of different gasifiers has been developed and marketed, all types geared towards handling the specific properties of a typical fuel or range of fuels. However, the universal gasifier that is able to handle all or most of the available fuels or fuel types, does not exist, and may not exist in the foreseeable future (Speight, 2008, 2013a, 2013b).

In fact, compared to a typical fossil fuel, the complex ligno-cellulosic structure of certain types of biomass is more difficult to react and gasify. The nature of the mineral impurities in biomass in conjunction with the presence of various inorganic species, as well as sulfur and nitrogen containing compounds, adversely impacts the thermal processing of the oxygenated hydrocarbonaceous structure of the biomass. In contrast to combustion of biomass feedstocks in which fuel-bound nitrogen and sulfur are converted to nitrogen oxides (NOx) and sulfur oxides (SOx), steam gasification involves thermal treatment under a reducing atmosphere resulting in fuel-bound nitrogen release as molecular nitrogen and fuel-bound sulfur conversion to hydrogen sulfide (H_2S) that is more easily removed by means of selected adsorbents (Mokhatab et al., 2006; Speight, 2009, 2013a, 2014).

Unlike combustion, the gasification process is more energy intensive. Careful engineering of the process reactor is necessary if the result is to produce rather than consume a significant amount of energy or power as a result of the thermal treatment. Thus, a careful selection must be made of the type of gasifier, with the recognition that the operation of the gasifier will be feedstock dependent.

Four types of gasifier configuration are currently available for commercial use: (1) fixed bed gasifier, which is subdivided into the counter-current fixed bed gasifier and the co-current fixed bed gasifier, (2) the fluid bed gasifier, (3) the entrained flow gasifier, and (4) processes involving the use of molten salt(s) or molten metal(s) (Speight, 2011, 2013a). All systems show relative advantages and disadvantages with respect to operation, and for this reason each will have its own technical and/or economic advantages in a specific set of circumstances.

However, each type of gasifier may be designed to operate either at atmospheric pressure or at high pressure. In the latter type of operation, hydrogasification of the feedstock (Chapter 2) is optimized and the quality of the product gas (in terms of heat, or Btu, content) is improved. In addition, the reactor size may be reduced and the need to pressurize the gas before it is introduced into a pipeline is eliminated (if a high-heat-content gas is to be the ultimate product). High pressure systems may have problems associated with the introduction of the feedstock into the reactor. Furthermore, low pressure or atmospheric pressure gasification reactors are frequently designed with an accompanying fuel gas compressor after the synthesis gas cleanup processes.

Furthermore some gasifiers are also categorized as *single stage units* or *multi-stage units* and the multi-stage units are further subcategorized into *single line operation* or *double-line operation*.

In *single stage* gasifiers, conversion takes place in a single reactor using steam, air, or oxygen. The common *single stage* technologies are fixed bed, fluidized bed, and entrained flow reactors. In fixed bed gasifiers the fuel is fed from the top and the gasification medium is injected at the bottom. The fuel moves slowly down the reactor with the gas moving upwards in a countercurrent direction whereas in co-current designs the fuel and the gasification agent move in the same direction and the fuel must pass successively through the drying, pyrolysis, oxidation, and reduction zones. The advantages of countercurrent designs are fewer restrictions on fuel moisture and particle size, no special fuel preparation is required, and a wide range of fuels can be used. By comparison, co-current gasifiers produce a better quality gas but place strict requirements on fuel properties. Fluid bed gasifiers allow for more efficient gasification due to elimination of hot spots in the reactor. They are suitable for a wide range of feedstocks and can be scaled up to relatively large plants. They are more expensive to build and need better gas cleaning due to high particulate content in the product gas. Fluidized beds have no defined reaction zones and the conversion of fuel and secondary pyrolytic reactions take place in the same volume. Tar conversion can be supported by introducing catalytically active bed materials.

Multi-stage gasifiers take into account the fuel conversion steps of drying, devolatilization, gasification, and combustion zones to enhance process efficiency and product gas quality by influencing and optimizing the operating parameters. These concepts are categorized as *single-line operation* in which the feedstock stream is passed through reactors arranged in series or *double-line operation* in which the mass main stream is divided into at least two partial streams which pass through reactors that are arranged in parallel. Furthermore, in the *double-line* process, combustion and gasification reactors are separate and are only connected by heat transfer — a pyrolysis stage is used to split the fuel into char and gas and to provide the heat necessary for operation; the char or part of the pyrolysis gas must be oxidized outside the pyrolysis reactor.

If chosen judiciously to accommodate the varying properties of the feedstock (Chapters 1 and 3), the gasifier will operate satisfactorily with respect to gas quality, efficiency, and pressure variations within certain ranges of the fuel properties of which the most important are: (1) feedstock reactivity, (2) feedstock size and size distribution, (3) bulk density of the feedstock, and (4) feedstock reactivity and propensity for char formation, (5) feedstock energy content, (6) feedstock moisture content, (7) volatile matter production, (7) mineral matter content, which is an indication of ash forming propensity, (8) ash chemical composition, and (9) ash reactivity. However, before choosing a gasifier that is theoretically suitable to the fuel it is important to consider the opposing view insofar as the fuel meeting the requirements of the gasifier or, failing that, whether or not the fuel can be treated to meet the necessary gasifier requirements.

2.1 Fixed Bed Gasifier

In a fixed bed gasifier, the feedstock is supported by a grate and combustion gases (such as steam, air, oxygen) pass through the supported feedstock after which the hot product gases exit from the top of the reactor. Heat is supplied internally or from an outside source, but some carbonaceous feedstocks (such as caking coal) cannot be used in an unmodified fixed bed reactor. Because of the operation of the gasifier, the system may also be referred to as a *descending-bed reactor* and is also often referred to as a *moving bed reactor* or, on occasion, a *countercurrent descending-bed reactor*.

In the gasifier, the feedstock (approximately 1/8 to 1 inch diameter or 3 to 25 mm diameter) is laid down at the top of a vessel while reactant gases are introduced at the bottom of the vessel and flow at relatively low velocity upward through the interstices between the feedstock particles. As the feedstock descends, the reactions are carried out in a countercurrent fashion: the first reaction is *devolatilization* by the sensible heat from the rising gas, then *gasification* of the resulting hydrogen-deficient char (Chapter 2). Hydrogenation by the hydrogen in the reactant gas may also occur but, eventually, the feedstock is converted to gases and mineral ash.

Thus, the *countercurrent fixed bed gasifier* (*updraft gasifier, counterflow gasifier*) consists of a fixed bed of carbonaceous fuel through which the gasification agent (steam, oxygen, and/or air) flows in countercurrent configuration. The ash is either removed dry or as a slag. Slagging gasifiers require a higher ratio of steam and oxygen to carbon in order to reach temperatures higher than the ash fusion temperature. The nature of the gasifier means that the fuel must have high mechanical strength and must be non-caking (especially when coal is used as a co-gasification feedstock) so that it will form a permeable bed, although recent developments may reduce these restrictions to some extent. The throughput for this type of gasifier is relatively low but the thermal efficiency is high as the exit gas temperature is relatively low and, as a result of the relatively low temperature, the production of methane and tar is significant.

The main advantage of this gasifier is the effective heat exchange in the reactor. High temperature synthesis dries the feedstock (an important aspect of gasification of wet biomass) as it moves downwards in the reactor. By the time heat exchange has taken place, the raw synthesis gas is cooled significantly on its way through the bed. The temperature of the synthesis at the reactor exit point is approximately 250°C (480°F — in a downdraft gasifier the exit gas temperature is on the order of 800°C, 1470°F). Since synthesis gas is exploited in order to dry the incoming feedstock, the system sensitivity to feedstock moisture content is less than in other gasification reactors. On the other hand, the countercurrent flow of feedstock and synthesis gas results in higher tar content (10 to 20% w/w) in the raw synthesis gas. Advantages of updraft gasification include: (1) relatively simple, low cost process, (2) equipped to process feedstocks (such as biomass) with a high moisture and high inorganic content (such as municipal solid waste), and (3) proven technology (Chopra and Jain, 2007).

The *co-current fixed bed (downdraft)* gasifier is similar to the countercurrent gasifier, but the gasification agent gas flows in co-current configuration with the fuel (downwards, hence the name downdraft gasifier); it is suitable for biomass fuels (Dogru et al., 2002; Zainal et al., 2002; Dogru, 2004; Tinaut et al., 2008; Wei et al., 2009). Heat needs to be added to the upper part of the bed, either by combusting small amounts of the fuel or from external heat sources. The product gas leaves the gasifier at a high temperature, and most of this heat is often transferred to the gasification agent added in the top of the bed, resulting in energy efficiency almost equivalent to the countercurrent gasifier. In this configuration, any tar produced must pass through a hot bed of char thereby converting much of the tar to gaseous product. As a result of tar conversion, tar concentration in the raw synthesis gas is less than in the case of updraft gasifiers.

The co-current fixed bed (downdraft) gasifier is easier to control than the countercurrent fixed bed gasifier but is more sensitive to the quality of the feedstock. For example, in the case of biomass feedstocks, while updraft gasifiers can process biomass with moisture content up to 50% w/w, in downdraft gasification a moisture content range between 10 and 25% w/w is required. The advantages of downdraft gasification are: (1) up to 99.9% w/w of the tar formed is consumed, requiring minimal or tar cleanup removal from the product gases, (2) feedstock minerals remain with the char/ash, reducing the need for a cyclone, and (3) relatively simple and low cost process. However, the disadvantages of downdraft gasification are: (1) the feedstock should be dried to a low moisture content (<20% w/w moisture), (2) the synthesis gas exiting the reactor is at high temperature, requiring a secondary heat recovery system, and (3) approximately 4 to 7% w/w of the carbon may remain unconverted.

Crossdraft gasification reactors, which operate well on dry air blast and dry fuel, do have advantages over updraft gasification reactors and downdraft gasifiers but the disadvantages — such as high exit gas temperature, poor carbon dioxide reduction and high gas velocity — which are the consequences of the design, often outweigh the advantages for many feedstocks.

Unlike downdraft and updraft gasifiers, the ash bin, fire, and reduction zone in crossdraft gasifiers are separated. This design tends to limit the type of fuel for operation to low mineral matter fuels such as wood, charcoal, and coke. The load following ability of the crossdraft gasifier is quite good due to concentrated partial zones which operate at temperatures up to 2000°C (3600°F). The relatively higher temperature in the crossdraft gasification reactor has an effect on gas composition, resulting in high carbon monoxide content and low hydrogen and methane content when dry fuel such as char or charcoal is used.

2.2 Fluid Bed Gasifier

In the *fluidized bed gasifier* (*fluid bed gasifier*), the fuel is fluidized in oxygen (or air) and steam and the ash is removed dry or as heavy agglomerates. The temperatures are relatively low in dry ash gasifiers, so the fuel must be highly reactive. Feedstock throughput is higher than for the fixed bed, but not as high as for the entrained flow gasifier. The conversion efficiency is variable and a recycle operation or subsequent combustion of solids may be necessary to increase conversion. Fluidized bed gasifiers are most useful for fuels that form highly corrosive ash (such as biomass) that would damage the walls of slagging gasifiers.

The fluidized bed system requires the feedstock to be finely ground into small particles and the reactant gases are introduced through a perforated deck near the bottom of the vessel. The volume rate of gas flow is such that its velocity is sufficient to suspend the solids but not high enough to blow them out of the top of the vessel. The result is a bed of solids which simulate a boiling action ensuing intimate contact with the upward flowing gas, leading to a uniform temperature distribution. The solid flows rapidly and repeatedly from bottom to top and back again, while the gas flows rather uniformly upward. As a result, reaction rates are faster than in the moving bed because of the intimate contact between gas and solids and the increased solids surface area due to the smaller particle size. Although no countercurrent flow is possible, a degree of countercurrent flow can be accomplished by placing two or more fluid bed stages one above the other.

Another aspect of *staged gasification* is to send the feedstock, such as biomass, to a pyrolysis reactor, where organic vapors are produced and sent to a gasifier to be reformed into a clean fuel gas (Leijenhorst and Van de Beld, 2009). In the top section of the gasifier, the vapors are mixed with (preheated) air to increase the temperature to 800 to 950°C (1470 to 1740°F). The bottom part can be filled with a reforming catalyst to convert remaining tar and ammonia. In the last stage the gas is cooled to ambient temperature.

Compared with the fixed bed gasifiers, the sequence of reactor processes (drying, pyrolysis, oxidation, and reduction) is not obvious at any specific points of the gasifier since these processes occur in the entire reactor thus resulting in a more homogeneous type of reaction, leading to more constant and lower temperatures inside the reactor and, thus, no *hot spots*. Due to the lower operating temperatures, ash does not melt and it is more easily removed from the reactor. In addition, sulfur-containing and chlorine-containing constituents of the feedstock can be absorbed in the inert bed material thus eliminating the fouling hazard and reducing the maintenance costs. Another significant difference is that fluidized bed gasifiers are much less subject to biomass quality (and mixed feedstocks) than fixed bed systems.

One critical advantage of a fluidized bed gasification system (as opposed to downdraft or fixed bed system) is the use of multiple feed stocks without experiencing downtime (Capareda, 2011). Another important characteristic of the fluidized bed system is the ability to operate at various throughputs without having to use a larger diameter unit. This is accomplished by changing the appropriate bed material. By using a larger bed material, a higher air flow rate is required for fluidization and thus more biomass may need to be fed at higher rates to maintain the same fuel to air ratio as before. The reactor free-board must be high enough so that bed materials are not blown out of the system. Also, a fluidized bed gasification reactor is designed to be accompanied by a cyclone downstream of the gasifier to capture the larger particles that are entrained out of the reactor as a result of the fluidity of the bed and the velocity of the gas rising through the bed. These particles are recycled back into the reactor. Overall, the residence time of feedstock particles in a fluidized bed gasifier is shorter than that of a moving bed gasifier.

Finally, depending on the inflow speed, the fluidized bed gasifier can be characterized either as a *bubbling fluidized bed* system or as a *circulating fluidized bed* system − the circulating fluidized bed system corresponds to higher velocity of the gasification medium.

A bubbling fluidized bed consists of fine, inert particles of sand or alumina, which are selected based on their suitability of physical properties such as size, density, and thermal characteristics. The gas flow rate is chosen to maintain the bed in a fluidization condition; gas enters at the bottom of the vessel. The dimension of the bed at some height above the distributor plate is increased to reduce the superficial gas velocity below the fluidization velocity to maintain inventory of solids and to act as a disengaging zone. A cyclone is used to trap the smaller size particles that exit the fluidized bed either to return fines to the bed or to remove ash rich fines from the system. The feedstock, such as biomass, is introduced either through a feed chute to the top of the bed or deep inside the bed. The deeper introduction of the feedstock into the bed of inert solids provides sufficient residence time for fines that would otherwise be entrained in the fluidizing gas. The organic constituent of the feedstock pyrolytically vaporize and are partially combusted in the bed. The exothermic combustion provides the heat to maintain the bed at temperature and to volatilize additional biomass.

A bubbling fluid bed design is generally more sensitive to bed utilization. The size of the feedstock particles greatly affects the rate of gasification and the ability of the biomass to migrate to the center of the bed in a bubbling fluid bed design. With small particles, gasification is rapid and unburned material might not make it to the center of the bed, resulting in oxygen slip and a void center in the bubbling fluid bed reactor. If all or a majority of the feedstock quickly gasifies, there will be insufficient char to maintain a uniform bed. For this reason, a bubbling fluid bed will generally require additional feed points that must be balanced for larger particle sizes.

The advantages of the bubbling fluidized bed gasifier are: (1) it yields a uniform product gas, (2) it exhibits a nearly uniform temperature distribution throughout the reactor, (3) it is able to accept a wide range of fuel particle sizes, including fines, (4) it provides high rates of heat transfer between inert material, fuel, and gas, and (5) high conversion is possible with low tar and unconverted carbon. The disadvantages of bubbling fluidized bed gasification are that a large bubble size may result in gas bypass through the fluid bed.

If the gas flow of a bubbling fluidized bed is increased, the gas bubbles become larger forming large voids in the bed and entraining substantial amounts of solids. The bubbles basically disappear in a circulating fluid bed in which the turbulent bed solids are collected, separated from the gas and returned to the bed, forming a solids circulation loop. A circulating fluid bed can be differentiated from a bubbling fluid bed in that there is no distinct separation between the dense solids zone and the dilute solids zone. Lower bed density can be achieved with an increase in gas flow rates in excess of transport velocity of the fluidized bed particles. The residence time of the solids in the circulating fluid bed is determined by the solids circulation rate, attrition of the solids, and the collection efficiency of the solids in the cyclones.

On the other hand, a circulating fluid bed design operates at a higher velocity and incorporates recycling of the char and bed material, resulting in complete mixing regardless of feedstock size. Generally, the circulating fluid bed designs are more flexible but are still limited by the amount of very fine material that can be processed. The advantages of the circulating fluidized bed gasifier are: (1) it is suitable for rapid reactions, (2) high heat transport rates are possible due to high heat capacity of the bed material, and (3) high conversion rates are possible with low tar and unconverted carbon. The disadvantages of the circulating fluidized bed gasifier are: (1) temperature gradients occur in the direction of solid flow, (2) the size of fuel particles determines minimum transport velocity — high velocities may result in equipment erosion, and (3) heat exchange is less efficient than with a bubbling fluidized bed (Babu, 2006).

A novel reactor design that is particularly appropriate for biomass is the *indirectly heated gasification* technology which utilizes a bed of hot particles (sand), which is fluidized using steam. Solids (sand and char) are separated from the synthesis gas via a cyclone and then transported to a second fluidized bed reactor. The second bed is air blown and acts as a char combustor, generating a flue gas exhaust stream and a stream of hot particles. The hot (sand) particles are separated from the flue gas and recirculated to the gasifier to provide the heat required for pyrolysis. The product gas is practically nitrogen-free and has a heating value of on the order of 400 Btu/ft^3 (Paisley et al., 1999; Turn, 1999).

Another novel design is the new fluidized bed gasifier with increased gas − solid interaction combining two circulating fluidized bed reactors (Schmid et al., 2011). The aim of the design is to generate a nitrogen-free product gas with low tar content and low fines (particulate matter) content. The system accomplishes this by division into an air/combustion and a fuel/gasification reactor − the two reactors are interconnected via loop seals to assure the global circulation of bed material.

The fuel/gasification reactor is a circulating fluidized bed but with the special characteristic of almost countercurrent flow conditions for gas phase and solids. The gas velocity and the geometrical properties in the fuel/gasification reactor are chosen in such a way that entrainment of coarse particles is low at the top. Due to the dispersed downward movement of the solids, volatile products are <u>not produced</u> in the upper part of the fuel reactor and the issues relating to insufficient gas phase conversion and high tar content are avoided.

Finally, effective mixing of feedstocks of various sizes is needed to maintain uniform temperature and a good mix depends on the relative concentrations of the solids in the bed and the velocity of the gas (Bilbao et al., 1988; Ghaly et al., 1989). Variations in the size, shape, and density of the fuel particles can have an adverse effect on mixing, which results in: (1) changes in temperature gradients within the reactor, (2) an increase in tar formation and agglomeration, and (3) a decrease in the efficiency of feedstock conversion (Cranfield, 1978; Bilbao et al., 1988).

2.3 Entrained Bed Gasifier

In the *entrained bed gasifier* (*entrained flow gasifier*) a dry pulverized solid, an atomized liquid fuel, or a fuel slurry is gasified with oxygen (air is a much less frequent gasification agent) in co-current flow and the gasification reactions take place in a dense cloud of very fine particles. The high temperatures and pressures also mean that a higher throughput can be achieved but the thermal efficiency is somewhat lower as the gas must be cooled before it can be cleaned with existing technology. The high temperatures also mean that tar and methane are not present in the product gas; however the oxygen requirement is higher than for the other types of gasifiers.

The design of an entrained flow reactor gives a residence time of the feedstock in the reaction zone to be on the order of seconds, or tens of seconds, which requires that the gasifier operate at high temperatures to achieve high carbon conversion. Consequently, most entrained flow gasifiers are designed to use oxygen rather than air and also to operate above the slagging temperature of the feedstock mineral matter.

The entrained flow reactor requires a smaller particle size of the feedstock than the fluid bed gasifier so that the feedstock can be conveyed pneumatically by the reactant gases − typically the fuel must be pulverized. In this case, there is little or no mixing of the solids and gases, except when the gas initially meets the solids. Furthermore, apart from the higher temperature, entrained flow gasification usually takes place at elevated pressure (pressurized entrained flow gasifiers) reaching operating pressures even up to 750 psi. The existence of such high temperatures and pressures requires more sophisticated reactor design and construction materials.

All entrained flow gasifiers are designed to remove the major part of the ash as a slag − the operating temperature of the gasifier is well above the ash fusion temperature. A smaller fraction of the ash is produced either as a very fine dry fly ash or as black colored fly ash slurry. Some fuels, in particular certain types of biomass, can form slag that is corrosive for ceramic inner walls that serve to protect the gasifier outer wall. However some entrained bed type of gasifiers do not possess a ceramic inner wall but have an inner water or steam cooled wall covered with partially solidified slag. For fuel that produces ash with a high ash fusion temperature, limestone can be mixed with the fuel prior to gasification in order to lower the ash fusion temperature.

2.4 Molten Salt Gasifier

The molten salt gasifier (molten metal gasifier) uses, as the name implies, a molten medium of an inorganic salt (or molten metal) to generate the heat to decompose the feedstock into products. There are a number of applications of molten bath gasification.

In the gasifiers, crushed feedstock, steam air, and/or oxygen are injected into a bath of molten salt, iron, or feedstock ash. The feedstock appears to *dissolve* in the melt where the volatiles crack and are converted into carbon monoxide and hydrogen. The feedstock carbon reacts with oxygen and steam to produce carbon monoxide and hydrogen. Unreacted carbon and mineral ash float on the surface from which they are discharged.

High temperatures are required to maintain a molten bath – approximately 900°C, 1650°F and above, depending on the nature of the melt. Such temperature levels favor high reaction rates and throughputs and low residence times. Consequently, tar and volatile oil products are not produced in any great quantity, if at all. Gasification may be enhanced by the catalytic properties of the melt used. Molten salts, which are generally less corrosive and have lower melting points than molten metals, can strongly catalyze the steam – coal reaction and lead to very high conversion efficiencies.

In the process, the carbonaceous feedstock devolatilizes with some thermal cracking of the volatile constituents leaving the char (sometime referred to as *fixed carbon*) and sulfur to dissolve in the molten salt (such as an iron salt) whereupon carbon is oxidized to carbon monoxide by oxygen introduced through lances placed at a shallow depth in the bath). The sulfur migrates from the molten salt to the slag layer where it reacts with lime to produce calcium sulfide.

The product gas, which leaves the gasifier at high temperature (usually approximately 1425°C, 2600°F), is cooled, compressed, and fed to a shift converter where a portion of the carbon monoxide is reacted with steam to attain a carbon monoxide to hydrogen ratio of 1:3 (Chapter 2). Any carbon dioxide produced is removed and the gas is again cooled and enters a methanator where carbon monoxide and hydrogen react to form methane (Chapter 2). Excess water is removed from the methane-rich product and − depending on the type of feedstock used and the extent of purification required − the final gas product may have a heat content of 920 Btu/ft^3.

As another example, the Pullman − Kellogg process involves contacting feedstock with a melt of an inorganic salt such as sodium carbonate. In the process, air is bubbled into the bottom of the gasifier through multiple inlet nozzles and the feedstock (typically sized to 1/4 in., 6 mm) is fed beneath the surface of the molten salt bath using a central feed tube whereupon natural circulation and agitation of the melt disperses the material. The main gasification reaction is a partial oxidation reaction and any volatile matter from the feedstock reacts to produce a fuel gas free of oils and tars, as well as ammonia. A water − gas shift equilibrium exists above the melt and, accordingly, in the reducing environment, carbon dioxide and water concentrations are minimal.

In practice, the molten salt design allows for some of the catalysis processes to take place within the gasifier instead of downstream. For example, if the reactor or process design allows the hydrogen and carbon monoxide to be produced in separate distinct streams, the need for postprocess separation prior to catalyzing into synthetic fuels will be eliminated. The molten salt/metal design is such that the use of a fluxing material, such as lime or limestone, is required. When combined with the silica ash that is generated through normal gasification, the slag produced and removed from the molten metal reactor can be used directly as cement or formed into bricks for construction materials.

2.5 Plasma Gasifier

Plasma is a high temperature, highly ionized (electrically charged) gas capable of conducting electrical current. Plasma technology has a long history of development and has evolved into a valuable tool for engineers and scientists who need to use very high temperatures for new process applications (Messerle and Ustimenko, 2007; Arena, 2012). Man-made plasma is formed by passing an electrical discharge through a gas such as air or oxygen (O_2). The interaction of the gas with the electric arc dissociates the gas into electrons and ions, and causes its temperature to increase significantly, often (in theory) exceeding 6000°C (10 830°F).

Serious efforts have been made over the last two decades, with some success, to apply plasma gasification technology to gasification technology and to treat industrial and municipal solid wastes (MSW). It is believed that the technology can be used as a gasification reactor thereby allowing: (1) greater feedstock flexibility enabling a variety of fuels such as coal, biomass, and municipal solid waste to be used as fuel without the need for pulverizing, (2) air blowing and thus an oxygen plant is not required, (3) high conversion (>99%) of carbonaceous matter to synthesis gas, (3) the absence of tar in the synthesis, (4) production of high-heating-value synthesis gas suitable for use in a combustion turbine operation, (5) production of little or no char, ash, or residual carbon, (6) production of a glassy slag with beneficial value, (7) high thermal efficiency, and (8) low carbon dioxide emissions.

In the process, the gasifier is heated by a plasma torch system located near the bottom of the reactor vessel. In the gasifier, the feedstock is charged into a vertical reactor vessel (refractory lined or water-cooled) at atmospheric pressure. A superheated blast of air, which may be enriched with oxygen, is provided to the bottom of the gasifier, at the stoichiometric amount required for gasification. The amount of air fed is such that the superficial velocity of the upward flowing gas is low, and that the pulverized feedstock can be fed directly into the reactor. Additional air and/or steam can be provided at different levels of the gasifier to assist with pyrolysis and gasification. The temperature of the synthesis gas leaving the top of the gasifier is maintained above 1000°C (1830°F). At this temperature, tar formation is eliminated.

Gasification takes place at very high temperatures, driven by the plasma torch system, which is located at the bottom of the gasifier vessel. The high operating temperatures break down the feedstock and/or all hazardous and toxic components into their respective elemental constituents, and dramatically increases the kinetics of the various reactions occurring in the gasification zone, converting all organic materials into hydrogen (H_2) and carbon monoxide (CO). Any residual materials from inorganic constituents of the feedstock (including heavy metals) will be melted and produced as a vitrified slag which is highly resistant to leaching.

2.6 Other Types

The *rotary kiln gasifier* is used in several applications, varying from industrial waste to cement production and the reactor accomplishes two objectives simultaneously: (1) moving solids into and out of a high temperature reaction zone and (2) assuring thorough mixing of the solids during reaction. The kiln is typically comprised of a steel cylindrical shell lined with abrasion-resistant refractory – to prevent overheating of the metal – and is usually inclined slightly toward the discharge port. The movement of the solids being processed is controlled by the speed of rotation of the kiln.

The moving grate gasifier is based on the system used for waste combustion in a waste-to-energy process. The constant-flow grate feeds the waste feedstock continuously to the incinerator furnace and provides movement of the waste bed and ash residue toward the discharge end of the grate. During the operation stoking and mixing of the burning material enhances distribution of the feedstocks and, hence, equalization of the feedstock composition in the gasifier. The thermal conversion takes place in two stages: (1) the primary chamber for gasification of the waste (typically at an equivalence ratio of 0.5) and (2) the secondary chamber for high temperature oxidation of the synthesis gas produced in the primary chamber (Grimshaw and Lago, 2010; Hankalin et al., 2011).

3 ENERGY BALANCE AND OTHER DESIGN OPTIONS

Fuels for gasification reactors differ significantly in chemical properties, physical properties, and morphological properties and, hence, require different reactor design and operation. It is for this reason that, during more than a century of gasification experience, a large number of different gasifiers has been developed – each reactor designed to accommodate the specific properties of a typical fuel or range of fuels. In short, the gasification reactor that is designed to accommodate all (or most) types of fuels does not exist.

However, before choosing a gasifier for any individual fuel it is important to ensure that the fuel meets the requirements of the gasifier or that it can be treated to meet these requirements. Practical tests are needed if the fuel has not previously been successfully gasified. In other words the fuel must match the gasifier and the gasifier must match the fuel.

The gasification reactor must be configured to accommodate the energy balance of the chemical reactions. During the gasification process, most of the energy bound up in the fuel is not released as heat. In fact, the fraction of the feedstock's chemical energy, or heating value, which remains in the product gases (especially the synthesis gas) is an important measure of the efficiency of a gasification process (which is dependent upon the reactor configuration) and is known as the *cold gas efficiency*. Most commercial-scale gasification reactors have a cold gas efficiency on the order of 65% to 80%, or even higher.

Thus, it is important for the reactor to limit the amount of heat that is transferred out of the zone where the gasification reactions are occurring. If not, the temperature within the gasification zone could be too low to allow the reactions to proceed – as an example, a minimum temperature on the order of 1000°C (1830°F) is typically needed to gasify coal. As a result, a gasification reactor is typically refractory lined with no water cooling to ensure as little heat loss as possible. Gasification reactors also typically operate at elevated pressure (often as high as 900 psia), which allows them to have very compact construction with minimum surface area and minimal heat loss.

In addition to being designed and selected for feedstock type, another design option for the gasification reactor involves the method for cooling the synthesis gas produced by the gasifier.

Regardless of the type of gasifier, the exiting synthesis gas must be cooled down to approximately 100°C (212°F) in order to utilize conventional acid gas removal technology. This can be accomplished either by passing the synthesis gas through a series of heat exchangers which recover the sensible heat for use (for example, in the stem cycle an *integrated combined cycle unit*, IGCC unit) or by directly contacting the synthesis gas with relatively cool water (a *quench* operation). The quench operation results in some of the quench water being vaporized and mixed with the synthesis gas. The quenched synthesis gas is saturated with water and must pass through a series of condensing heat exchanges which remove the moisture from the synthesis gas (so it can be recycled to the quench zone).

Quench designs have a negative impact on the heating rate of related equipment (such as the IGCC unit) because the sensible heat of the high temperature synthesis gas is converted to low level process heat rather than high pressure steam. However, quench designs have much lower capital costs and can be justified when low cost feedstock (such as biomass or waste) is available. Quench designs also have an advantage if carbon dioxide capture is desired. The saturated synthesis gas exiting a quench section has near the optimum water/carbon monoxide ratio as the feedstock to a water − gas shift reactor which will convert the carbon monoxide to carbon dioxide. Non-quench designs that require carbon dioxide capture need to add steam to the synthesis gas before it is sent to a water − gas shift reactor.

4 CHEMICAL ASPECTS

Generally, gasification involves two distinct stages that are both feed-stock and reactor dependent: (1) devolatilization to produce a semi-char which is converted to char by elimination of hydrogen once the rate of devolatilization has passed a maximum, followed by (2) gasifi-cation of the char, which is specific to the reactor and the conditions of the reaction.

4.1 Feedstock Quality

The physical process parameters and feedstock type have an influence on gasification. For example, the reactivity of coal generally decreases with increase in rank (from lignite to sub-bituminous coal to bitumi-nous coal anthracite). Furthermore, the smaller the particle size, the more contact area between the coal and the reaction gases, leading to a more rapid reaction. For medium rank coal and low rank coal, reac-tivity increases with an increase in pore volume and surface area, but for coal having a carbon content greater than 85% w/w these factors have no effect on reactivity. In fact, in high rank coal, pore sizes are so small that the reaction is diffusion controlled.

Other feedstocks (such as petroleum residua and biomass) are so variable that gasification behavior and products vary over a wide range. The volatile matter produced during the thermal reactions varies widely and the ease with which tar products are formed as part of the gaseous products makes gas cleanup more difficult.

The mineral matter content of the feedstock also has an impact on the composition of the produced synthesis gas. Gasifiers may be designed to remove the produced ash in solid or liquid (slag) form. In fluidized or fixed bed gasifiers, the ash is typically removed as a solid, which limits operational temperatures in the gasifier to well below the ash melting point. In other designs, particularly slagging gasifiers, the operational temperatures are designed to be above the ash melting temperature. The selection of the most appropriate gasifier is often dependent on the melting temperature and/or the softening temperature of the ash and the feedstock which is to be used at the facility.

High moisture content of the feedstock lowers internal gasifier temperatures through evaporation and the endothermic reaction of steam and char. Usually a limit is set on the moisture content of feedstock supplied to the gasifier, which can be met by drying operations if necessary. For a typical fixed bed gasifier and moderate carbon content and mineral matter content of the feedstock, the moisture limit may be on the order of 35% w/w. Fluidized bed and entrained bed gasifiers have a lower tolerance for moisture, limiting the moisture content to approximately 5 to 10% w/w of the feedstock. Oxygen supplied to the gasifiers must be increased with an increase in mineral matter content (ash production) or moisture content in the feedstock.

Depending on the type of feedstock being processed and the analysis of the gas product desired, pressure also plays a role in product definition (Speight, 2011, 2013a). In fact, some (or all) of the following processing steps will be required: (1) pretreatment of the feedstock, (2) primary gasification, (3) secondary gasification of the carbonaceous residue – char – from the primary gasifier, (4) removal of carbon dioxide, hydrogen sulfide, and other acid gases, (5) shift conversion for adjustment of the carbon monoxide/hydrogen mole ratio to the desired ratio, and (6) catalytic methanation of the carbon monoxide/hydrogen mixture to form methane. If high-heat-content (high-Btu) gas is desired, all of these processing steps are required since gasifiers do not yield methane in the concentrations required (Speight, 2008, 2011, 2013a).

Thus, the reactivity of the feedstock is an important factor in determining the design of the reactor because feedstock reactivity, which determines the rate of reduction of carbon dioxide to carbon monoxide in the reactor, influences reactor design insofar as it dictates the height needed in the reduction zone.

In addition certain operational design characteristics of the reactor system (load following response, restarting after temporary shutdown) are affected by the reactivity of the char produced in the reactor. There is also a relationship between feedstock reactivity and the number of active places on the char surface, these being influenced by the morphological characteristics as well as the geological age of the fuel. The grain size and the porosity of the char produced in the reduction zone influence the surface available for reduction and, therefore, the rate of the reduction reactions which are facilitated by reactor design.

4.2 Mixed Feedstocks

Both fixed bed and fluidized bed gasifiers have been used in co-gasification of coal and biomass – these include a downdraft fixed bed gasifier (Kumabe et al., 2007; Speight, 2011). However, operational problems experienced when a fluidized bed gasifier was employed included (1) defluidization of the fluidized bed gasifier caused by agglomeration of low melting point ash present in the biomass, and (2) clogging of the downstream pipes due to excessive tar accumulation (Pan et al., 2000; Vélez et al., 2009). In addition, co-gasification and co-pyrolysis of birch wood and coal in an updraft fixed bed gasifier as well as in a fluidized bed gasifier has yielded overhead products with 4.0 to 6.0% w/w tar content while the fixed bed reactor gave tar yields on the order of 25 to 26% w/w for co-gasification of coal and silver birch wood mixtures (1:1 w/w ratio) at 1000°C (1830°F) (Collot et al., 1999).

From the perspective of the efficient operation of the reactor, the presence of mineral matter has a deleterious effect on fluidized bed reactors. The low melting point of ash formed from the mineral matter present in woody biomass can lead to agglomeration which influences the efficiency of the fluidization – the ash can cause sintering, deposition, and corrosion of the gasifier construction metal. In addition, biomass containing alkali oxides and salts can cause clinkering/slagging problems (McKendry, 2002).

4.3 Feedstock Devolatilization

The devolatilization (or pyrolysis) process commences at approximately 200 to 300°C (390 to 570° F), depending upon the nature and properties of the feedstock (Chapter 2). Volatile products are released and a carbonaceous residue (char) is produced, resulting in up to 70% weight loss for many feedstocks. The process determines the structure and composition of the char, which will then undergo gasification reactions.

More specifically, as the feedstock particle is heated, any residual moisture (assuming that the feedstock has been pre-dried) is driven off and after all the moisture contained in the feedstock particle(s) has evaporated, the particles undergo devolatilization. The devolatilization and discharge of volatiles generates a range of products varying from carbon monoxide and methane to high molecular weight hydrocarbons comprising paraffin/olefin hydrocarbons, aromatic hydrocarbons, heavy oil, and tar, which are also feedstock dependent. As these products pass from the devolatilization (pyrolysis) zone further thermal reactions will occur and gasification of the volatile products will commence.

At temperatures above 500°C (930°F) the conversion of the feedstock to char and mineral matter ash is completed. The gasification of char particles occurs after the devolatilization process has finished (Silaen and Wang, 2008). For gas generation the char provides the necessary energy to promote further heating and, typically, the char is contacted with air or oxygen and steam to generate the product gases.

For some feedstocks, carbon conversion is believed to be independent of the devolatilization rate and less sensible to feedstock particle size, but it is sensitive to the heterogeneous char − oxygen, char − CO_2, and char − steam reaction kinetics (Chen et al., 2000).

4.4 Char Gasification

The *gasification* process occurs as the char reacts with gases such as carbon dioxide and steam to produce carbon monoxide and hydrogen (Chapter 2). Also, corrosive ash elements such as chloride and potassium may be refined out by the gasification process, allowing the high temperature combustion of the gas from otherwise problematic feedstocks.

Although the initial gasification stage is completed in seconds or even less at elevated temperature, the subsequent gasification of the char produced at the initial gasification stage is much slower, requiring minutes or hours to obtain significant conversion under practical conditions and reactor designs for commercial gasifiers are largely dependent on the reactivity of the char, which in turn depends on the nature of feedstock. The reactivity of char also depends upon parameters of the thermal process required to produce the char from the original feedstock. The rate of gasification of the char decreases as the process temperature increases due to the decrease in active surface area of char. Therefore a change of char preparation temperature may change the chemical nature of char, which in turn may change the gasification. The reactivity of char may be influenced by catalytic effect of mineral matter in the char.

Heat and mass transfer processes in fixed or moving bed gasifiers are affected by complex solids flow and chemical reactions. Moving bed gasifiers are countercurrent flow reactors in which the feedstock enters at the top of the reactor and oxygen (air) enters at the bottom of the reactor (Beenackers, 1999). Because of the countercurrent flow arrangement of the reactor, the heat of reaction from the gasification reactions serves to pre-heat the coal before it enters the gasification reaction zone. Consequently, the temperature of the synthesis gas exiting the gasifier is significantly lower than the temperature needed for complete conversion of the feedstock. However, coarsely crushed feedstock may settle while undergoing (1) thermal drying, (2) pyrolysis-devolatilization, (3) gasification, and (4) reduction. In addition, the particles change in diameter, shape, and porosity − non-ideal behavior may result from bridges, gas bubbles, channeling, and a variable void fraction may also change heat and mass transfer characteristics.

Though there is a considerable overlap of the processes, each can be assumed to occupy a separate zone where fundamentally different chemical and thermal reactions take place. The gasification technology package consists of a fuel and ash handling system and gasification system − reactor, gas cooling, and cleaning system. There are also auxiliary systems namely, the water treatment plant to meet the requirements of industry and pollution control boards. The prime mover for power generation consists of either a diesel engine or a spark ignited engine coupled to an alternator. In the case of thermal systems, the end use device is a standard industrial burner.

Depending on the gasifier technology employed and the operating conditions, significant quantities of water, carbon dioxide, and methane can be present in the product gas, as well as a number of minor and trace components. Under the reducing conditions in the gasifier, most of the sulfur in the fuel sulfur is converted to hydrogen sulfide (H_2S) as well as to smaller yields of carbonyl sulfide (COS). Organically bound nitrogen in the feedstock is generally (but not always) converted to gaseous nitrogen (N_2) – some ammonia (NH_3) and a small amount of hydrogen cyanide (HCN) are also formed. Any chlorine in the feedstock (such as coal) is converted to hydrogen chloride (HCl) with some chlorine present in the particulate matter (fly ash). Trace elements, such as mercury and arsenic, are released during gasification and partition among the different phases, such as fly ash, bottom ash, slag, and product gas.

4.5 Mineral Matter Content and Ash Production

Finally, gasification reactors are very susceptible to ash production and properties. Ash can cause a variety of problems particularly in up or downdraught gasifiers. Slagging or clinker formation in the reactor, caused by melting and agglomeration of ashes, at best will greatly add to the difficulty of gasifier operation. If no special measures are taken, slagging can lead to excessive tar formation and/or complete blocking of the reactor. A worst case scenario is the possibility of air-channeling which can lead to a risk of explosion, especially in updraught gasifiers.

Whether or not slagging occurs depends on the ash content of the fuel, the melting characteristics of the ash, and the temperature pattern allowed by gasifier design. Local high temperatures in voids in the fuel bed in the oxidation zone, caused by bridging in the bed, may cause slagging even when using fuels with a high ash melting temperature.

Generally, slagging is not observed with fuels having mineral matter ash contents less than 5 to 6% w/w. Severe slagging can be expected for fuels having mineral matter contents in excess of 12% w/w. For fuels with mineral matter contents between 6 and 12%, the slagging behavior depends to a large extent on the mineral matter composition – reflected in the ash melting temperature – which is influenced by the presence of trace elements giving rise to the formation of low melting point eutectic mixtures.

Updraught and downdraught gasification reactors are able to operate with slagging fuels if specially modified (continuously moving grates and/or external pyrolysis gas combustion). Cross-draught gasification reactors, which work at temperatures on the order of 1500°C (2700°F) and higher, need special safeguards with respect to the mineral matter content of the fuel. Fluidized bed reactors, because of their inherent capacity to control the operating temperature, suffer less from ash melting and fusion problems.

5 GASIFIER — FEEDSTOCK COMPATIBILITY

All gasifier designs show relative advantages and disadvantages with respect to feedstock type, application and simplicity of operation, and for this reason each gasifier — feedstock relationship will have specific technical and/or economic advantages in a particular set of circumstances.

However, before selecting a feedstock for a gasifier it is important to ensure that it is compatible with the requirements of the gasifier or that it can be treated to meet these requirements. Furthermore, a series of test methods may need to be applied to the feedstock if the analysis is unknown or speculative and to determine the gasifier-feedstock compatibility. Thus, each type of gasifier will operate satisfactorily with respect to feedstock character, gas quality, and process efficiency only within certain ranges of the feedstock properties of which the most important are: (1) feedstock reactivity, (2) energy content, (3) moisture content, (4) production of volatile matter, (5) particle size and distribution, (6) bulk density, (7) propensity for char formation, (8) mineral matter content, and (9) ash yield.

5.1 Feedstock Reactivity

Feedstock reactivity is an important factor determining the rate of reduction of carbon dioxide to carbon monoxide in a gasifier. Reactivity influences the reactor design insofar as it dictates the height needed in the reduction zone — fluidized bed gasifiers show great promise in gasifying a number of agricultural wastes.

Furthermore, co-gasification of various feedstocks (such as coal and biomass) may be advantageous from a chemical point of view — some practical problems have been associated with upstream, gasification, and downstream processes. On the upstream side, the particle size of the feedstock is required to be uniform for optimum gasification. In addition, moisture content and pretreatment (torrefaction) are very important during upstream processing.

In addition, most wood species have ash production yields that are less than 2% w/w of the feedstock and are therefore suitable fixed bed gasifiers. However, because of the highly volatile content of wood, updraught systems produce a tar-containing gas suitable mainly for direct burning. Cleaning of the gas to make it suitable for engines is rather difficult and capital and labor intensive. Downdraught systems can be designed to deliver a virtually tar-free product gas in a certain capacity range when fuelled by wood blocks or wood chips of low moisture content. However, most currently available downdraught gasifiers axe not suitable for non-pelletized sawdust. Problems encountered are: (1) excessive tar production, (2) inadmissible pressure drop, and (3) lack of bunker flow. On the other hand, fluidized bed gasifiers can accommodate small sawdust particles and produce burner quality gas.

In principle, many countries (especially developing countries) have a wide range of agricultural residues available for gasification but in practice, however, experience with most types of waste is extremely limited. Coconut shells and maize cobs are the best documented and seem unlikely to create serious problems in fixed bed gasifiers. Coconut husks can give rise to bridging problems in the bunker section, but the material can be gasified when mixed with a certain quantity of wood. Most cereal straws have ash contents above 10% and present slagging problems in downdraught gasifiers – in fact rice husks can produce ash on the order of 20% w/w of the feedstock and, because of this, may be the most difficult feedstock for gasification.

It is possible to gasify most types of agricultural waste in updraught gasifiers. However, the capital, maintenance and labor costs, and the environmental consequences (disposal of tarry condensates) involved in cleaning the gas prevent engine malfunction under most circumstances. Downdraught equipment is cheaper to install and operate and creates fewer environmental difficulties, but at present technology is inadequate to handle agricultural residues (with the possible exception of maize cobs and coconut shells) without installing expensive (and partly unproven) additional devices.

In addition certain operational characteristics of the gasification system (load following response, restarting after temporary shutdown) are affected by the reactivity of the char produced in the gasifier. Reactivity is dependent on the type of feedstock — feedstocks such as wood, charcoal, and peat are far more reactive than coal.

After the initial reaction in the gasifier, the reactivity of the char becomes important since there is a relation between reactivity and the number of active sites on the char surface, these being influenced by the morphological characteristics of the char and the feedstock from which the char was produced. The grain size and the porosity of the char produced in the reduction zone influence the surface available for reduction and, therefore, the rate of the reduction reactions.

Another aspect of the properties of the char is the effect of various elements that act as catalysts on the rate of gasification. Small quantities of potassium, sodium, and zinc can have a large effect on the reactivity of the fuel.

5.2 Energy Content

The choice of a fuel for gasification will in part be decided by its heating value. The method of measurement of the fuel energy content will influence the estimate of efficiency of a given gasification system. Reporting of fuel heating values is often confusing since at least three different bases are used: (1) higher heating values as obtained in an adiabatic bomb calorimeter — these values include the heat of condensation of the water that is produced during combustion and because it is very difficult to recover the heat of condensation in actual gasification operations these values present a too optimistic view of the fuel energy content; (2) higher heating values on a moisture-free basis, which disregard the actual moisture content of the fuel and so provide even more optimistic estimates of energy content; (3) higher heating values on a moisture and ash free basis, which disregard the incombustible components and consequently provide estimates of energy content too high for a given weight of fuel, especially in the case of some agricultural residues (rice husks).

The only realistic way therefore of presenting feedstock heating values for gasification purposes is to give lower heating values (excluding the heat of condensation of the water produced) on an ash inclusive basis and with specific reference to the actual moisture content of the fuel.

Plastics waste, being a potential energy source is another possible feedstock for fluid bed gasifiers (Mastellone and Arena, 2007). Gasification of plastics can be subdivided into the following sequence of steps: (1) heating and melting of polymer particles, (2) primary cracking of polymer chain with consequent formation of intermediate hydrocarbon fragments, and (3) secondary cracking of intermediates with formation of methane, hydrogen, olefins, and oxidation/reduction reactions with the formation of carbon monoxide, carbon dioxide, and water. Ternary reactions can also occur with the subsequent formation of aromatic products and, in the presence of metals, coke.

A suitable method to avoid or reduce tar formation during fluidized bed gasification is the catalytic removal of tar precursors and intermediates. In particular, cycloparaffins, naphthenes, and aromatics, forming during ternary reactions of the intermediate species produced by primary cracking, can be decomposed to carbon and hydrogen by means of metal-based catalysts. These contain transition metals such as iron, cobalt, nickel, chromium, vanadium platinum, and magnesium, i.e., those metals typically used for the reforming of hydrocarbons (Wu and Williams, 2010).

5.3 Moisture Content

The heating value of the gas produced by any type of gasifier depends at least in part on the moisture content of the feedstock (Chapter 1). Moisture content can be determined on a dry basis as well as on a wet basis. In this chapter the moisture content on a dry basis will be used.

A high moisture content of the fuel reduces the thermal efficiency since heat is used to drive off the water and consequently this energy is not available for the reduction reactions and for converting thermal energy into chemical bound energy in the gas. Therefore, high moisture contents result in low gas heating values. When the gas is used for direct combustion purposes, low heating values can be tolerated and the use of feedstocks with a moisture content (dry basis) of up to 40 to 50% w/w is feasible, especially when using updraught gasifiers.

In downdraught gasifiers high moisture contents give rise not only to low gas heating values, but also to low temperatures in the oxidation zone, and this can lead to insufficient tar converting capability if the gas is used for engine applications. Both because of the gas heating value and issues related to tar entrainment, downdraught gasifiers need reasonably dry fuels (less than 25% w/w moisture dry basis).

5.4 Production of Volatile Matter
The amount of volatile matter produced from the feedstock determines the necessity of special measures (either in design of the gasifier or in the layout of the gas cleanup train) in order to remove tars from the product gas in engine applications. In practice the only biomass fuel that does not need this special attention is good-quality charcoal.

The volatile matter produced by charcoal, however, is often underestimated and in practice may be anything from 3 to 30% w/w or more. As a general rule if the fuel has the ability to produce more than 10% w/w volatile matter it should be used in downdraught gas producers, but even in this case the method of charcoal production should be taken into account. Charcoal produced in large scale retorts is fairly consistent in volatile matter content, but large differences can be observed in charcoal produced from small scale open pits or portable metal kilos that are common in most developing countries.

5.5 Particle Size and Distribution
Many feedstocks require drying and size reduction before they can be fed into a gasifier. Size reduction is needed to obtain appropriate particle sizes; however, drying is required to achieve moisture content suitable for gasification operations. In addition, densification of biomass may be done to make pellets and improve density and material flow in the feeder areas.

Up- and downdraught gasifiers are limited in the range of fuel size acceptable in the feed stock. Fine grained and/or fluffy feedstock may cause flow problems in the bunker section of the gasifier as well as an inadmissible pressure drop over the reduction zone and a high proportion of dust in the gas. Large pressure drops will lead to reduction of the gas load of downdraught equipment, resulting in low temperatures and tar production.

Excessively large sizes of particles or pieces give rise to a reduced reactivity of the fuel, resulting in startup problems and poor gas quality, and to transport problems through the equipment. A large range in size distribution of the feedstock will generally aggravate the above phenomena. The presence of large-sized particles can cause gas channeling, especially in updraught gasifiers.

Acceptable sizes of the feedstocks for gasification systems depend to a certain extent on the design of the units. In general, wood gasifiers operate on wood blocks and woodchips ranging from $8 \times 4 \times 4$ cm. to $1 \times 0.5 \times 0.5$ cm. Charcoal gasifiers are generally fuelled by charcoal lumps ranging between $1 \times 1 \times 1$ cm and $3 \times 3 \times 3$ cm. Fluidized bed gasifiers are normally able to handle fuels with particle diameters varying between 0.1 and 20 mm.

5.6 Bulk Density
The bulk density of a feedstock is the weight per unit volume of loosely packed feedstock and feedstocks with a high bulk density are advantageous because they represent a high energy-for-volume value. Consequently these fuels need less bunker space for a given refueling time. Feedstocks with a low bulk density can give rise to insufficient flow under gravity, resulting in low gas heating values and ultimately in burning of the char in the reduction zone – inadequate bulk densities can be improved by briquetting or pelletizing.

5.7 Propensity for Char Formation
The occurrence of physical and morphological difficulties with charcoal produced in the oxidation zone has been reported. Some feedstocks (especially softwoods) produce char that shows a tendency to disintegrate. In extreme cases this may lead to inadmissible pressure drop.

A number of tropical hardwoods (notably teak) are reported to call for long residence times in the pyrolysis zone, leading to bunker flow problems, low gas quality, and tar entrainment.

5.8 Mineral Matter Content
The presence of mineral matter in the coal-biomass feedstock is not appropriate for fluidized bed gasification. Low melting point of ash present in woody biomass leads to agglomeration which causes defluidization of the ash and sintering and deposition as well as corrosion of the gasifier construction metal bed (Vélez et al., 2009). Biomass-containing alkali oxides and salts are likely to produce clinkering/slagging problems from ash formation (McKendry, 2002). It is imperative

to be aware of the melting of biomass ash, its chemistry within the gasification bed (no bed, silica/sand, or calcium bed), and the fate of alkali metals when using fluidized bed gasifiers.

The occurrence of slagging in the gasifier depends on: (1) the ash produced from the fuel, (2) the melting characteristics of the ash, and (3) the temperature pattern in the gasifier. Local high temperatures in voids in the fuel bed in the oxidation zone, caused by bridging in the bed, may cause slagging even when using fuels with a high ash melting temperature.

5.9 Ash Yield

In general, no slagging is observed with fuels having ash production less than 6% w/w of the feedstock but severe slagging can be expected for feedstocks where the mineral matter content is higher than 12% w/w. For feedstocks with ash production from 6 to 12% w/w of the feedstock, the propensity for slagging depends on the ash melting temperature, which is influenced by the presence of trace elements giving rise to the formation of low melting point eutectic mixtures. For gasification purposes the melting behavior of the fuel ash should be determined in both oxidizing and reducing atmospheres.

Updraught and downdraught gasifiers are able to operate with slagging fuels if specially modified (continuously moving grates and/or external pyrolysis gas combustion). Cross-draught gasifiers, which work at very high temperatures of 1500°C (2730° F) and above, need special safeguards with respect to the ash formation from the feedstock. Fluidized bed reactors, because of their inherent capacity to control the operating temperature, suffer less from ash melting and fusion problems.

6 PRODUCTS

Gasification agents are typically air, oxygen-enriched air, or oxygen and the products of the combustion or gasification oxidation reaction change significantly as the oxygen-to-fuel ratio changes from combustion to gasification conditions (Table 3.2), which are dependent upon gasifier design and operation.

The mixture under gasifying conditions is fuel-rich and there is not enough oxygen to effect complete conversion of the feedstock, in terms

Table 3.2 Comparison of Products from Combustion and Gasification Processes

	Combustion	Gasification
Carbon	CO_2	CO
Hydrogen	H_2O	H_2
Nitrogen	NO, NO_2	HCN, NH_3 or N_2
Sulfur	SO_2 or SO_3	H_2S or COS
Water	H_2O	H_2

of gas quality. As a result, the feedstock carbon reacts to produce carbon instead of carbon dioxide and the feedstock hydrogen is converted to hydrogen rather than to water. Thus, the quantity and quality of the gas generated in a gasification reactor is influenced not only by the feedstock characteristics but predominantly by the gasifier type and configuration as well as by the amount of air, oxygen, or steam introduced into the system, which is also influenced by the gasifier configuration.

At the same time, the fate of the nitrogen and sulfur in the fuel is also dictated by oxygen availability (i.e., the configuration of the gasification reactor). The presence of nitrogen and sulfur in a gasification process has important environmental consequences. Instead of being converted to the respective oxides, the fuel-bound nitrogen is predominantly converted to molecular nitrogen (N_2) and hydrogen cyanide (HCN) while the sulfur in the fuel produces hydrogen cyanide (HCN) and carbonyl sulfide (COS).

Steam is sometimes added for temperature control, heating value enhancement, or to permit the use of external heat (*allothermal gasification*). The major chemical reactions break and oxidize hydrocarbons to give a product gas containing carbon monoxide (CO), carbon dioxide (CO_2), hydrogen (H_2), and water (H_2O). Other important components include hydrogen sulfide (H_2S), various compounds of sulfur and carbon, ammonia, low molecular weight hydrocarbons, and tar.

As a very general *rule of thumb*, optimum gas yields and gas quality are obtained at operating temperatures of approximately 595 to 650° C (1100 to 1200° F). A gaseous product with a higher heat content (BTU/ft^3) can be obtained at lower system temperatures but the overall yield of gas (determined as the *fuel-to-gas ratio*) is reduced by the unburned portion of the feedstock, which usually appears as char.

Furthermore, introduction of the water – gas equilibrium concept provides the opportunity to calculate the gas composition theoretically from a gasifier which has reached equilibrium at a given temperature. The procedure is to derive from mass balances of the four main ingoing elements (carbon, hydrogen, oxygen, and nitrogen) an energy balance over the system and the relationship given by the water – gas equilibrium. By further assuming that the amounts of methane in the producer gas per weight of dry fuel are constant (as is more or less the case with gasifiers under normal operating conditions) a set of relationships becomes available permitting the calculation of gas compositions for a wide range of input parameters (feedstock moisture content) and system characteristics (heat losses through convection, radiation, and sensible heat in the gas).

6.1 Primary Products

The products from gasification may be of low, medium, or high heat content (high-Btu) as dictated by the process as well as by the ultimate use for the gas (Chapter 2) (Speight, 2008, 2011, 2013a, 2013b).

Product gases from fixed bed versus fluidized bed gasifier configurations vary significantly. Fixed bed gasifiers are relatively easy to design and operate and are best suited for small to medium-scale applications with thermal requirements of up to several megawatts thermal (MWt). For large scale applications, fixed bed gasifiers may encounter problems with bridging of the feedstock (especially in the case of biomass feedstocks) and non-uniform bed temperatures. Bridging leads to uneven gas flow, while non-uniform bed temperature may lead to hot spots, ash formation, and slagging. Large scale applications are also susceptible to temperature variations throughout the gasifier because of poor mixing in the reaction zone.

Pressurized gasification systems lend themselves to economical synthesis gas production and can also be more flexible in production turndown depending on the reactor design. Typically this is the case for both a pressurized bubbling reactor and a circulating fluidized bed reactor, while the flexibility of an atmospheric fluidized bed reactor is typically limited to narrower pressure and production ranges. Both designs are well suited for pressurized synthesis gas production. Pressurized designs require more costly reactors, but the downstream equipment (such as gas cleanup equipment, heat exchangers, synthesis gas reactors) will consist of fewer and less expensive components (Worley and Yale, 2012).

6.2 Secondary Gaseous Products

There is a series of products that are called by older (even archaic) names that should also be mentioned here as clarification. Since these products are not the desired synthesis products and/or hydrogen, they are labeled here (for convenience) as secondary gaseous products.

Producer gas is a low-Btu gas typically obtained from a coal gasifier (fixed bed) upon introduction of air instead of oxygen into the fuel bed. The composition of the producer gas is approximately 28% v/v carbon monoxide, 55% v/v nitrogen, 12% v/v hydrogen, and 5% v/v methane with some carbon dioxide.

Water-gas is a medium-Btu gas that is produced by the introduction of steam into the hot fuel bed of the gasifier. The composition of the gas is approximately 50% v/v hydrogen and 40% v/v carbon monoxide with small amounts of nitrogen and carbon dioxide.

Town gas is a medium-Btu gas that is produced in the coke ovens and has the approximate composition 55% v/v hydrogen, 27% v/v methane, 6% v/v carbon monoxide, 10% v/v nitrogen, and 2% v/v carbon dioxide. Carbon monoxide can be removed from the gas by catalytic treatment with steam to produce carbon dioxide and hydrogen.

Synthetic natural gas (SNG) is methane obtained from the reaction of carbon monoxide or carbon with hydrogen. Depending on the methane concentration, the heating value can be in the range of high-Btu gases.

6.3 Tar

Another key contribution to efficient gasifier operation is the need for a tar reformer. Tar reforming occurs when water vapor in the incoming synthesis gas is heated to a sufficient temperature to cause steam reforming in the gas conditioning reactor, converting condensable hydrocarbons (tars) to non-condensable lower molecular weight products (Vreugdenhil and Zwart, 2009; Arena, 2012). The residence time in the conditioning reactor is sufficient to allow a water − gas shift reaction to occur also and generate increased amounts of hydrogen in the synthesis gas.

Thus, tar reforming technologies — which can be thermally driven and/or catalytically driven — are utilized to break down or <u>decompose tar products</u> and high boiling hydrocarbon products into hydrogen and carbon monoxide. This reaction increases the H_2/CO ratio of the synthesis gas and reduces or eliminates tar condensation in downstream process equipment. Thermal tar reformer designs are typically fluid bed or fixed bed type. Catalytic tar reformers are filled with heated loose catalyst material or catalyst block material and can be fixed or fluid bed designs.

Typically, the tar reformer is a refractory lined steel vessel equipped with catalyst blocks, which may contain a noble metal or a nickel enhanced material. Synthesis gas is routed to the top of the vessel and flows down through the catalyst blocks. Oxygen and steam are added to the tar reformer at several locations along the flow path to enhance the synthesis gas composition and achieve optimum performance in the reformer. The tar reformer utilizes a catalyst to decompose tars and heavy hydrocarbons into hydrogen and carbon monoxide. Without this decomposition the tars and heavy hydrocarbons in the synthesis gas will condense as the synthesis gas is cooled in the downstream process equipment. In addition, the tar reformer increases the hydrogen/carbon monoxide ratio for optimal conversion. The synthesis gas is routed from the tar reformer to downstream heat recovery and gas cleanup unit

REFERENCES

Arena, U., 2012. Process and technological aspects of municipal solid waste gasification. Rev. Waste Manage. 32, 625–639.

Babu, S.P., 2006. Thermal gasification of biomass. Proceedings. Workshop No. 1: Perspectives on Biomass Gasification. IEA Bioenergy Agreement, Task 33. International Energy Agency, Paris, France.

Beenackers, A.A.C.M., 1999. Biomass gasification in moving beds: a review of european technologies. Renewable Energy 16, 1180–1186.

Bilbao, R., Lezaun, J.L., Menendez, M., Abanades, J.C., 1988. Model of mixing/segregation for sand – straw mixtures in fluidized beds. Powder Technol. 56, 149–151.

Capareda, S., 2011. Advances in gasification and pyrolysis research using various biomass feedstocks. Proceedings. 2011 Beltwide Cotton Conferences, Atlanta, Georgia, January 4 – 7, pp. 467–472.

Chen, C., Horio, M., Kojima, T., 2000. Numerical simulation of entrained flow coal gasifiers. Part II: effects of operating conditions on gasifier performance. Chem. Eng. Sci. 55 (18), 3875–3883.

Chopra, S., Jain, A., 2007. A review of fixed bed gasification systems for biomass. Agricultural Engineering International: CIGR Ejournal. Invited Overview No. 5. vol. IX. April.

Collot, A.G., 2002. Matching Gasifiers to Coals. Report No. CCC/65. Clean Coal Centre, International Energy Agency, London, UK.

Collot, A.G., 2006. Matching gasification technologies to coal properties. Int. J. Coal Geol. 65, 191–212.

Collot, A.G., Zhuo, Y., Dugwell, D.R., Kandiyoti, R., 1999. Co-pyrolysis and cogasification of coal and biomass in bench-scale fixed-bed and fluidized bed reactors. Fuel 78, 667–679.

Cranfield, R., 1978. Solids mixing in fluidized beds of large particles. AIChE J. 74 (176), 54–59.

Dogru, M., 2004. Gasification of leather residues-part I. Experimental study via a pilot scale air blown downdraft gasifier. Energy Sources Part A 26, 35–44.

Dogru, M., Howarth, C.R., Akay, G., Keskinler, B., Malik, A.A., 2002. Gasification of hazelnut shells in a downdraft gasifier. Energy 27, 415–427.

Ghaly, A.E., Al-Taweel, A.M., Hamdullahpur, F., Ugwu, I., 1989. Physical and chemical properties of cereal straw as related to thermochemical conversion. Proceedings. 7th. Bioenergy R&D Seminar. E.N. Hogan (Editor). Ministry of Energy, Mines, and Resources Ministry, Ottawa, Ontario, Canada, pp. 655–661.

Grimshaw, A.J., Lago, A., 2010. Small scale energos gasification technology. Proceedings of the Third International Symposium on Energy from Biomass and Waste. Venice, Italy. November 8 – 11. CISA Publishers, Padova, Italy.

Hankalin, V., Helanti, V., Isaksson, J., 2011. High efficiency power production by gasification. Proceedings. 13th International Waste Management and Landfill Symposium. S. Margherita di Pula, Cagliari, Italy. October 3 – 7. CISA Publishers, Padova, Italy.

Kumabe, K., Hanaoka, T., Fujimoto, S., Minowa, T., Sakanishi, K., 2007. Cogasification of woody biomass and coal with air and steam. Fuel 86, 684–689.

Leijenhorst, E., Van de Beld, L., 2009. Staged gasification: clean fuel through innovative coupling of existing thermochemical conversion systems. Proceedings 17th European Biomass Conference and Exhibition. Hamburg, Germany, June 29 – July 3, pp. 847–854.

Mastellone, M.L., Arena, U., 2007. Fluidized bed gasification of plastic waste: effect of bed material on process performance. Proceedings. 55th International Energy Agency – Fluidized Bed Conversion Meeting, Paris, France, October 30–31.

McKendry, P., 2002. Energy production from biomass part 3: gasification technologies. Bioresour. Technol. 83 (1), 55–63.

Messerle, V.E., Ustimenko, A.B., 2007. Solid Fuel Plasma Gasification. Advanced Combustion and Aerothermal Technologies. NATO Science for Peace and Security Series C. Environmental Security, pp. 141–1256.

Mokhatab, S., Poe, W.A., Speight, J.G., 2006. Handbook of Natural Gas Transmission and Processing. Elsevier, Amsterdam, Netherlands.

Paisley, M.A., Farris, M.C., Black, J., Irving, J.M., Overend, R.P., 1999. Commercial demonstration of the battelle/FERCO biomass gasification process: startup and initial operating experience. Proceedings of the Fourth Biomass Conference of the Americas, Oakland, California, August 29 – September 2. Elsevier Science Ltd., Oxford, UK.

Pan, Y.G., Velo, E., Roca, X., Manyà, J.J., Puigjaner, L., 2000. Fluidized-bed cogasification of residual biomass/poor coal blends for fuel gas production. Fuel 79, 1317–1326.

Schmid, J.C., Pfeifer, C., Kitzler, H., Pröll, T., Hofbauer, H., 2011. A new dual fluidized bed gasifier design for improved in situ conversion of hydrocarbons. Proceedings International Conference on Polygeneration Strategies (ICPS 2011), Vienna, Austria. August 30–September 1.

Silaen, A., Wang, T., 2008. Effects of turbulence and devolatilization models on gasification simulation. Proceedings 25th International Pittsburgh Coal Conference, Pittsburgh, Pennsylvania. September 29 – October 2.

Speight, J.G., 2008. Synthetic Fuels Handbook: Properties, Processes, and Performance. McGraw-Hill, New York.

Speight, J.G., 2009. Enhanced Recovery Methods for Heavy Oil and Tar Sands. Gulf Publishing Company, Houston, Texas.

Speight, J.G. (Ed.), 2011. The Biofuels Handbook. Royal Society of Chemistry, London, UK.

Speight, J.G., 2013a. The Chemistry and Technology of Coal, third ed. CRC Press, Taylor and Francis Group, Boca Raton, FL.

Speight, J.G., 2013b. Coal-Fired Power Generation Handbook. Scrivener Publishing, Salem, MA.

Speight, J.G., 2014. The Chemistry and Technology of Petroleum 5th Edition. CRC Press, Taylor and Francis Group, Boca Raton, FL.

Tinaut, F.V., Melgar, A., Pérez, J.F., Horillo, A., 2008. Effect of biomass particle size and air superficial velocity on the gasification process in a downdraft fixed bed gasifier. An experimental and modeling study. Fuel Process. Technol. 89, 1076–1089.

Turn, S.Q., 1999. Biomass integrated gasifier combined cycle technology: application in the cane sugar industry. Int. Sugar J. 101, 1205.

Vélez, J.F., Chejne, F., Valdés, C.F., Emery, E.J., Londoño, C.A., 2009. Cogasification of colombian coal and biomass in a fluidized bed: an experimental study. Fuel 88, 424–430.

Vreugdenhil, B.J., Zwart, R.W.R., 2009. Tar Formation in Pyrolysis and Gasification. Report No. ECN-E-08-087. Energy Research Center of the Netherlands, Petten, Netherlands.

Wei, L., Thomasson, J.A., Bricka, R.M., Sui, R., Wooten, J.R., Columbus, E.P., 2009. Syngas quality evaluation for biomass gasification with a downdraft gasifier. Trans. ASABE 52 (1), 21–37.

Worley, M., Yale, J., 2012. Biomass Gasification Technology Assessment. Subcontract Report No. NREL/SR-5100-57085. National Renewable Energy Laboratory, Golden, Colorado. November.

Wu, C., Williams, P.T., 2010. Pyrolysis-gasification of plastics, mixed plastics and real-world plastic waste with and without Ni-Mg-Al catalyst. Fuel 89, 3022–3032.

Zainal, Z.A., Rifau, A., Quadir, G.A., Seetharamu, K.N., 2002. Experimental investigation of a downdraft biomass gasifier. Biomass Bioenergy 23 (4), 283–289.

CHAPTER 4

Applications

1 INTRODUCTION

With the rapid increase in the use of gasification technology, particularly using coal as the feedstock, from the fifteenth century onwards (Nef, 1957; Taylor and Singer, 1957) it is not surprising that the concept of using coal to produce a flammable gas, especially the use of water and hot coal (van Heek and Muhlen, 1991), became commonplace (Elton, 1958). In fact, the production of gas from coal has been a vastly expanding area of coal technology, leading to numerous research and development programs. As a result, the characteristics of rank, mineral matter, particle size, and reaction conditions are all recognized as having a bearing on the outcome of the process; not only in terms of gas yields but also on gas properties (Massey, 1974; van Heek and Muhlen, 1991). The products from the gasification of coal may be of low-, medium-, or high-heat-content (high-Btu) as dictated by the process as well as by the ultimate use for the gas (Fryer and Speight, 1976; Mahajan and Walker, 1978; Anderson and Tillman, 1979; Cavagnaro, 1980; Bodle and Huebler, 1981; Argonne, 1990; Baker and Rodriguez, 1990; Probstein and Hicks, 1990; Lahaye and Ehrburger, 1991; Matsukata et al., 1992; Speight, 2013a).

The gasification of coal, biomass, petroleum, or any carbonaceous residues is typically focused on feedstock conversion to gaseous products. In fact, gasification offers one of the most versatile methods (with a reduced environmental impact with respect to combustion) to convert carbonaceous feedstocks into electricity, hydrogen, and other valuable energy products as well as a wide range of chemical products (Speight, 2013a, 2013b). Depending on the previously described type of gasifier (e.g., air-blown, enriched oxygen-blown) and the operating conditions, gasification can be used to produce a fuel gas that is suitable for several applications.

2 FUEL GASES

The products of gasification are varied insofar as the gas composition varies with the system employed (Speight, 2013a). It is emphasized that the gas product must be first freed from any pollutants such as particulate matter and sulfur compounds before further use, particularly when the intended use is a water – gas shift or methanation (Cusumano et al., 1978; Probstein and Hicks, 1990).

2.1 Synthesis Gas

Synthesis gas (*syngas*) is a mixture mainly of hydrogen and carbon monoxide which is comparable in its combustion efficiency to natural gas (Speight, 2007, 2008). This reduces the emissions of sulfur, nitrogen oxides, and mercury, resulting in a much cleaner fuel (Sondreal et al., 2004, 2006; Lee et al., 2007; Yang et al., 2007; Nordstrand et al., 2008; Wang et al., 2008; Lee and Shah, 2013). The resulting hydrogen gas can be used for electricity generation or as a transport fuel. The gasification process also facilitates capture of carbon dioxide emissions from the combustion effluent (see discussion of carbon capture and storage below).

Although synthesis gas can be used as a standalone fuel, the energy density of synthesis gas is approximately half that of natural gas and is therefore mostly suited for use in producing transportation fuels and other chemical products. Thus, synthesis gas is mainly used as an intermediary building block for the final production (synthesis) of various fuels such as synthetic natural gas, methanol and synthetic petroleum fuel (dimethyl ether – synthesized gasoline and diesel fuel) (Chadeesingh, 2011; Speight, 2013a).

The use of synthesis gas offers the opportunity to furnish a broad range of environmentally clean fuels and chemicals and there has been steady growth in the traditional uses of synthesis gas. Almost all hydrogen gas is manufactured from synthesis gas and there has been an increase in the demand for this basic chemical. In fact, the major use of synthesis gas is in the manufacture of hydrogen for a growing number of purposes, especially in petroleum refineries (Speight, 2014). Methanol not only remains the second largest consumer of synthesis gas but has shown remarkable growth as part of the methyl ethers used as octane enhancers in automotive fuels.

The Fischer – Tropsch synthesis remains the third largest consumer of synthesis gas, mostly for transportation fuels but also as a growing feedstock source for the manufacture of chemicals, including polymers. The hydroformylation of olefins (the Oxo reaction), a completely chemical use of synthesis gas, is the fourth largest use of carbon monoxide and hydrogen mixtures. A direct application of synthesis gas as fuel (and eventually also for chemicals) that promises to increase is its use for *integrated gasification combined cycle* (IGCC) units for the generation of electricity (and also chemicals) from coal, petroleum coke, or heavy residuals. Finally, synthesis gas is the principal source of carbon monoxide, which is used in an expanding list of carbonylation reactions, which are of major industrial interest.

2.2 Low-Heat-Content (Low-Btu) Gas

During the production of coal gas by oxidation with air, the oxygen is not separated from the air and, as a result, the gas product invariably has a low-heat-content (150 to 300 Btu/ft^3). Low-heat-content gas is also the usual product of in situ gasification of coal which is used essentially as a method for obtaining energy from coal without the necessity of mining the coal, especially if the coal cannot be mined or of mining is uneconomical.

Several important chemical reactions and a host of side reactions are involved in the manufacture of low-heat-content gas under the high temperature conditions employed (Chadeesingh, 2011; Speight, 2011a, 2013a). Low-heat-content gas contains several components, four of which are always major components present at levels of at least several percent; a fifth component, methane, is marginally a major component.

The nitrogen content of low-heat-content gas ranges from somewhat less than 33% v/v to slightly more than 50% v/v and cannot be removed by any reasonable means; the presence of nitrogen at these levels makes the product gas *low-heat-content* by definition. The nitrogen also strongly limits the applicability of the gas to chemical synthesis. Two other non-combustible components (water, H_2O, and carbon dioxide, CO) further lower the heating value of the gas; water can be removed by condensation and carbon dioxide by relatively straightforward chemical means.

The two major combustible components are hydrogen and carbon monoxide; the H_2/CO ratio varies from approximately 2:3 to about 3:2. Methane may also make an appreciable contribution to the heat-content of the gas. Of the minor components hydrogen sulfide is the most signifi-cant and the amount produced is, in fact, proportional to the sulfur content of the feed coal. Any hydrogen sulfide present must be removed by one, or more, of several procedures (Mokhatab et al., 2006; Speight, 2014).

Low-heat-content gas is of interest to industry as a fuel gas or even, on occasion, as a raw material from which ammonia, methanol, and other compounds may be synthesized.

2.3 Medium-Heat-Content (Medium-Btu) Gas

Medium-heat-content gas has a heating value in the range 300 to 550 Btu/ft^3 and the composition is much like that of low-heat-content gas, except that there is virtually no nitrogen. The primary combustible gases in medium-heat-content gas are hydrogen and carbon monoxide (Kasem, 1979). Medium-heat-content gas is considerably more versatile than low-heat-content gas; like low-heat-content gas, medium-heat-content gas may be used directly as a fuel to raise steam, or used through a combined power cycle to drive a gas turbine, with the hot exhaust gases employed to raise steam, but medium-heat-content gas is especially amenable to synthesize methane (by methanation), higher hydrocarbons (by Fischer − Tropsch synthesis), methanol, and a variety of synthetic chemicals.

The reactions used to produce medium-heat-content gas are the same as those employed for low-heat-content gas synthesis, the major difference being the application of a nitrogen barrier (such as the use of pure oxygen) to keep diluent nitrogen out of the system.

In medium-heat-content gas, the H_2/CO ratio varies from 2:3 to 3:1 and the increased heating value correlates with higher methane and hydrogen contents as well as with lower carbon dioxide contents. Furthermore, the very nature of the gasification process used to pro-duce the medium-heat-content gas has a marked effect upon the ease of subsequent processing. For example, the CO_2-acceptor product is quite amenable to use for methane production because it has: (1) the desired H_2/CO ratio just exceeding 3:1, (2) an initially high methane content, and (3) relatively low water and carbon dioxide contents. Other gases may require appreciable shift reaction and removal of large quantities of water and carbon dioxide prior to methanation.

2.4 High-Heat-Content (High-Btu) Gas

High-heat-content gas is essentially pure methane and often referred to as *synthetic natural gas* or *substitute natural gas* (SNG) (Kasem, 1979; Speight, 1990, 2013a). However, to qualify as substitute natural gas, a product must contain at least 95% methane giving an energy content (heat-content) of synthetic natural gas on the order of 980 to 1080 Btu/ft^3.

The commonly accepted approach to the synthesis of high-heat-content gas is the catalytic reaction of hydrogen and carbon monoxide:

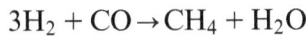

$$3H_2 + CO \rightarrow CH_4 + H_2O$$

To avoid catalyst poisoning, the feed gases for this reaction must be quite pure and, therefore, impurities in the product are rare. The large quantities of water produced are removed by condensation and recirculated as very pure water through the gasification system. The hydrogen is usually present in slight excess to ensure that the toxic carbon monoxide is reacted; this small quantity of hydrogen will lower the heat-content to a small degree.

The carbon monoxide/hydrogen reaction is somewhat inefficient as a means of producing methane because the reaction liberates large quantities of heat. In addition, the methanation catalyst is troublesome and prone to poisoning by sulfur compounds and the decomposition of metals can destroy the catalyst. Thus, hydrogasification may be employed to minimize the need for methanation:

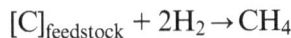

$$[C]_{feedstock} + 2H_2 \rightarrow CH_4$$

The product of hydrogasification is far from pure methane and additional methanation is required after hydrogen sulfide and other impurities are removed.

2.5 Substitute Natural Gas

Gasification can also be used to create substitute natural gas from coal by using a *methanation* reaction in which the coal-based synthesis gas — mostly carbon monoxide and hydrogen — can be converted to methane.

Substitute natural gas (*synthetic natural gas*) is an artificially produced version of natural gas which can be produced from coal, biomass, petroleum coke, or solid waste. The carbon-containing mass can be gasified and the resulting synthesis gas converted to methane, the major component of natural gas. There are several advantages associated with producing substitute natural gas. In times when natural gas is in short supply, substitute natural gas from coal could be a major driver for energy security by diversifying energy options and reducing imports of natural gas, thus helping to stabilize fuel prices.

Biomass and other low-cost feedstocks (such as municipal waste) can also be used along with coal to produce substitute natural gas. The use of biomass would reduce greenhouse gas emissions, as biomass is a carbon-neutral fuel. In addition, the development of substitute natural gas technology would also boost the other gasification-based technologies such as hydrogen generation and IGCC.

Identical to conventional natural gas (methane, CH_4), the resulting substitute natural gas can be transported in existing natural gas pipeline networks and used to generate electricity, produce chemicals/fertilizers, or heat homes and businesses. For many countries that lack natural gas resources, substitute natural gas enhances domestic fuel security by displacing imported natural gas that is likely to be supplied in the form of *liquefied natural gas* (LNG).

2.6 Hydrogen

The use of hydrogen in refining processes was perhaps the single most significant advance in refining technology during the 20th century and is now an inclusion in most refineries. In fact, a critical issue facing the refineries at present and in the future is the changing slate of crude oil feedstocks and the conversion of these feedstocks into refined transportation fuels under an environment of increasingly more stringent clean fuel regulations, decreasing heavy fuel oil demand, and increasing supply of heavy, high-sulfur crude oils. Hydrogen network optimization is at the forefront of world refineries options to address clean fuel trends, to meet growing transportation fuel demands, and to continue to make a profit from petroleum (Long et al., 2011). A key element of a hydrogen network analysis in a refinery involves the capture of hydrogen in its fuel streams and extending its flexibility and processing options. Thus, innovative hydrogen network optimization will be a critical factor influencing future refinery operating flexibility and profitability in a shifting world of crude feedstock supplies and ultra-low-sulfur (ULS) gasoline and diesel fuel.

Upgrading feedstocks such as heavy oils and residua evolved after the introduction of hydrodesulfurization processes (Speight and Ozum, 2002; Hsu and Robinson, 2006; Gary et al., 2007; Rana et al., 2007; Ancheyta and Speight, 2007). In the early days, the goal was desulfurization but, in later years, the processes were adapted to a 10 to 30% partial conversion operation, as intended to achieve desulfurization and obtain low-boiling fractions simultaneously, by increasing severity in operating conditions. However, as refineries have evolved and feedstocks have changed, refining heavy feedstocks has become a major issue in modern refinery practice and several process configurations have evolved to accommodate the heavy feedstocks (Khan and Patmore, 1997; Speight and Ozum, 2002; Hsu and Robinson, 2006; Gary et al., 2007; Speight, 2011a, 2014).

Hydrogen, one of the two major components of synthesis gas, is used to produce high-quality gasoline, diesel fuel, and jet fuel, meeting the requirements for clean fuels in state and federal clean air regulations (Ciferno and Marano, 2002). Hydrogen is also used to upgrade heavy crude oil and tar sand bitumen. Refineries can gasify low-value residuals, such as petroleum coke, asphalts, tars, and some oily wastes from the refining process to generate both the required hydrogen and the power and steam needed to run the refinery.

The hydrodesulfurization of the refined petroleum products requires pure hydrogen, which is usually obtained from the one contained in the oil itself. Crude oils with unfavorable sulfur to hydrogen ratio have a lower price on the market, but they require extra hydrogen for the desulfurization process. Hence, there is room for introducing an extra source of hydrogen from sources such as biomass and waste. Despite the technological challenges of such a process, there is the potential for an economic process from the purchase of low price crude oil with high sulfur content.

Thus, the gasification of petroleum residua, petroleum coke, and other feedstocks such as biomass (Agus et al., 2006; Speight, 2008, 2011a, 2011b, 2014) to produce hydrogen and/or power may become an attractive option for refiners. The premise that the gasification section of a refinery will be the *garbage can* for deasphalter residues, high-sulfur coke, as well as other refinery wastes is worthy of consideration. Other processes such as ammonia dissociation, steam − methanol interaction, or electrolysis are also available for hydrogen production, but economic factors and feedstock availability assist in the choice between processing alternatives.

2.7 Other Products

Hydrogen and carbon monoxide, the major components of synthesis gas, are the basic building blocks of a number of other products, such as fuels, chemicals, and fertilizers. In addition, a gasification plant can be designed to produce more than one product at a time (*co-production* or *polygeneration*), such as electricity and chemicals (e.g., methanol or ammonia).

2.7.1 Chemicals and Fertilizers

The process of producing energy using the gasification method has been in use for more than one hundred years. During that time coal was used to power these plants and, initially developed to produce town gas for lighting and heating in the 1800s, this was replaced by electricity and natural gas as well as use in blast furnaces. But the bigger role for gasification (of coal) was in the production of synthetic chemicals for which it has been in use since the 1920s. The concept is again being considered as a means of producing much-needed chemicals with the added benefit that not only coal but also low-value carbonaceous and/or hydrocarbonaceous feedstocks are gasified in a large chemical reactor. The resulting synthesis gas is cleansed and then converted into high value products such as synthetic fuels, chemicals, and fertilizers.

Typically, the chemical industry uses gasification to produce methanol as well as a variety of other chemicals – such as ammonia and urea – which form the foundation of nitrogen-based fertilizers and the production of a variety of plastics. The majority of the operating gasification plants worldwide are designed to produce chemicals and fertilizers.

3 LIQUID FUELS

The production of liquid fuels from coal via gasification is often referred to as the *indirect liquefaction* of coal (Speight, 2013a). In these processes, the coal is not converted directly into liquid products but involves a two-stage conversion operation in which coal is first converted (by reaction with steam and oxygen) to produce a gaseous mixture that is composed primarily of carbon monoxide and hydrogen (synthesis gas). The gas stream is subsequently purified (to remove sulfur, nitrogen, and any particulate matter) after which it is catalytically converted to a mixture of liquid hydrocarbon products.

3.1 Hydrocarbon Synthesis

The synthesis of hydrocarbons from carbon monoxide and hydrogen (synthesis gas) (the Fischer − Tropsch synthesis) is a procedure for the indirect liquefaction of coal and other carbonaceous feedstocks (Batchelder, 1962; Dry, 1976; Anderson, 1984; Speight, 2011a, 2011b). This process is the only coal liquefaction scheme currently in use on a relatively large commercial scale; South Africa is currently using the Fischer − Tropsch process on a commercial scale in their SASOL complex

Thus, coal is converted to gaseous products at temperatures in excess of 800°C (1470°F), and at moderate pressures, to produce synthesis gas:

$$[C]_{coal} + H_2O \rightarrow CO + H_2$$

The gasification may be attained by means of any one of several processes or even by gasification of coal in place (underground, or in situ, gasification of coal).

In practice, the Fischer − Tropsch reaction is carried out at temperatures of 200 to 350°C (390 to 660°F) and at pressures of 75 to 4000°psi. The hydrogen/carbon monoxide ratio is typically on the order of 2/2:1 or 2/5:1. Since up to three volumes of hydrogen may be required to achieve the next stage of the liquids production, the synthesis gas must then be converted by means of the water − gas shift reaction to the desired level of hydrogen:

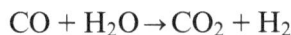

$$CO + H_2O \rightarrow CO_2 + H_2$$

After this, the gaseous mix is purified and converted to a wide variety of hydrocarbons:

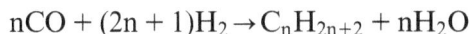

$$nCO + (2n + 1)H_2 \rightarrow C_nH_{2n+2} + nH_2O$$

These reactions result primarily in low-boiling and medium-boiling aliphatic compounds suitable for gasoline and diesel fuel.

Gasification is the foundation for converting coal and other solid feedstocks and natural gas into transportation fuels, such as gasoline, ultra-clean diesel fuel, jet fuel, naphtha, and synthetic oils. Two options are available for converting carbonaceous feedstocks to motor fuels via gasification.

In the *first option*, the synthesis gas undergoes an additional process, the Fischer − Tropsch reaction, to convert it to a liquid petroleum product. The Fischer − Tropsch process, with coal as a feedstock, was invented in the 1920s, was used by Germany during World War II, and has been utilized in South Africa for decades. Currently, it is also used in Malaysia and the Middle East with natural gas as the feedstock. In the *second option*, the *methanol-to-gasoline* process (MTG process), the synthesis gas is first converted to methanol (a commercially used process) and the methanol is then converted to gasoline by reacting it over catalysts.

The Fischer − Tropsch synthesis produces hydrocarbons of different chain lengths from a gaseous mixture of hydrogen and carbon monoxide. The higher molecular weight hydrocarbons can be hydrocracked to form mainly diesel of excellent quality. The fraction of short-chain hydrocarbons is used in a combined cycle plant with the remainder of the synthesis gas. The prognosis for the route to fuels and the use of gasification of biomass and conversion of the gaseous product(s) to Fischer − Tropsch-fuels in the transport sector is very promising. However, large-scale (pressurized) biomass gasification-based systems are necessary with particular attention given to the gas cleaning section.

3.2 Transportation Fuels from Tar Sand Bitumen

Tar sand deposits (oil sands deposits) can be found in many countries throughout the world, and may comprise more than 65% v/v of the total world oil reserve. The two largest deposits are in Canada and Venezuela − the tar sand deposits in Canada comprise three major deposits covering a region estimated to be more than 54 000 square miles (approximately 140 000 square kilometers). Estimates from the Alberta Energy and Utilities Board indicate that approximately 1.6 trillion barrels (1.6×10^{12} bbls) of crude oil equivalent are contained within the tar sand deposits of Canada. Of this amount, more than 170 billion barrels (170×10^{9} bbls) are considered recoverable, but this amount is dependent on current oil prices.

Gasification, a commercially proven technology that can be used to convert petroleum coke into synthesis gas, is being recognized as a means to economically generate hydrogen, power, and steam for tar sand operators in northeastern Alberta, Canada.

The tar sand deposits in Alberta (Canada) are estimated to contain as much recoverable bitumen as the petroleum available from the vast oil fields of Saudi Arabia. However, to convert the raw bitumen to saleable products requires extracting the bitumen from the sand and refining the separated bitumen to transportation fuels. The mining process requires massive amounts of steam to separate the bitumen from the sand and the refining process demands large quantities of hydrogen to upgrade the raw distillates to saleable products. Residual materials from the bitumen upgrading process include petroleum coke, deasphalted residua, and vacuum residua, all of which contain unused energy that can be released and captured for use by gasification. Traditionally, tar sand operators have utilized natural gas to produce the steam and hydrogen needed for the mining, upgrading, and refining processes.

Traditionally, oil sand operators have utilized natural gas to produce the steam and hydrogen needed for the mining, upgrading, and refining processes. However, a number of operators will soon gasify coke to supply the necessary steam and hydrogen. Not only will gasification displace expensive natural gas as a feedstock, it will also enable the extraction of useable energy from what is otherwise a very low-value product (coke). In addition, black water from the mining and refining processes can be recycled to the gasifiers using a wet feed system, reducing fresh water usage and waste water management costs — traditional oil sand operations consume large volumes of water.

4 POWER GENERATION

The gasification of any carbonaceous residue, such as coal, is the conversion of the feedstock to gaseous products (Ishi, 1982; Hotchkiss, 2003; Speight, 2013a). In fact, gasification offers one of the most versatile methods (with a lesser environmental impact than combustion) to convert carbonaceous feedstocks into electricity, hydrogen, and other valuable energy products.

Feedstock-to-power through gasification technology allows the continued use of domestic supplies of coal without the high level of air emissions associated with conventional coal-burning technologies. One of the advantages of the coal gasification technology is that it offers polygeneration: co-production of electric power, liquid fuels, chemicals, and hydrogen from the synthesis gas generated from gasification.

4.1 General Aspects

Depending on the type of gasifier (e.g., air-blown, enriched oxygen-blown) and the operating conditions, gasification can be used to produce a fuel gas that is suitable for several applications. Gasification for electric power generation enables the use of a technology common in modern gas fired power plants, the use of *combined cycle* technology to recover more of the energy released by burning the fuel. The use of these two types of turbines in the combined cycle system involves (1) a combustion turbine and (2) a steam turbine. The increased efficiency of the combined cycle for electrical power generation results in a 50% v/v decrease in carbon dioxide emissions compared to conventional coal plants. As the technology requires the development of economical methods of carbon*sequestration*, the removal of carbon dioxide from gaseous byproducts to prevent its release to the atmosphere, gasification units could be modified to further reduce their climate change impact because a large part of the carbon dioxide generated can be separated from the other product gas *before* combustion.

Gasification has been considered for many years as an alternative to combustion of solid or liquid fuels. It is easier to clean gaseous mixtures than it is to clean solid or high-viscosity liquid fuels. Clean gas can be used in internal combustion-based power plant that would suffer from severe fouling or corrosion if solid or low quality liquid fuels were burned inside them.

In fact, the hot synthesis gas that is produced by gasification of carbonaceous feedstocks can then be processed to remove sulfur compounds, mercury, and particulate matter before it is used to fuel a combustion turbine generator to produce electricity. The heat in the exhaust gases from the combustion turbine is recovered to generate additional steam. This steam, along with the steam produced by the gasification process, then drives a steam turbine generator to produce additional electricity. In the last decade, the primary application of gasification to power production has become more common due to the demand for high efficiency and low environmental impact.

As anticipated, the quality of the gas generated in a system is influenced by feedstock characteristics, gasifier configuration, and the amount of air, oxygen or steam introduced into the system. The output and quality of the gas produced is determined by the equilibrium established when the heat of oxidation (combustion) balances the heat of vaporization and volatilization plus the sensible heat (temperature rise) of the exhaust gases. The quality of the outlet gas (Btu/ft^3) is determined by the amount of volatile gases (such as hydrogen, carbon monoxide, water, carbon dioxide, and methane) in the gas stream. With some feedstocks, the higher the amounts of volatile produced in the early stages of the process the higher the heat-content of the product gas. In some cases, the highest gas quality may be produced at the lowest temperatures but when the temperature is too low, char oxidation reaction is suppressed and the overall heat-content of the product gas is diminished.

Gasification agents are normally air, oxygen-enriched air, or oxygen. Steam is sometimes added for temperature control, heating value enhancement, or to permit the use of external heat (*allothermal gasification*). The major chemical reactions break and oxidize hydrocarbons to give a product gas of carbon monoxide, carbon dioxide, hydrogen, and water. Other important components include hydrogen sulfide, various compounds of sulfur and carbon, ammonia, light hydrocarbons, and heavy hydrocarbons (tars).

Depending on the gasifier technology employed and the operating conditions, significant quantities of water, carbon dioxide, and methane can be present in the product gas, as well as a number of minor and trace components. Under the reducing conditions in the gasifier, most of the feedstock sulfur converts to hydrogen sulfide (H_2S), but $3 - 10\%$ converts to carbonyl sulfide (COS). Organically bound nitrogen in the coal feedstock is generally converted to gaseous nitrogen (N_2), but some ammonia (NH_3) and a small amount of hydrogen cyanide (HCN) are also formed. Any chlorine in the coal is converted to hydrogen chloride (HCl) with some chlorine present in the particulate matter (fly ash). Trace elements, such as mercury and arsenic, are released during gasification and partition among the different phases, such as fly ash, bottom ash, slag, and product gas.

Furthermore, an *integrated gasification combined cycle* power plant (IGCC power plant) combines the gasification process with a *combined cycle* power block (consisting of one or more gas turbines and a steam turbine). Clean synthesis gas is combusted in high efficiency gas turbines to produce electricity. The excess heat from the gas turbines and from the gasification reaction is then captured, converted into steam, and sent to a steam turbine to produce additional electricity.

In the IGCC power plant – where power generation is the focus – the clean synthesis gas is combusted in high efficiency gas turbines to generate electricity with very low emissions. The gas turbines used in these plants are slight modifications of proven, natural gas combined cycle gas turbines that have been specially adapted for use with synthesis gas. For IGCC power plants that include carbon dioxide capture, these gas turbines are adapted to operate on synthesis gas with higher levels of hydrogen. Although state-of-the-art gas turbines are commercially ready for the high-hydrogen synthesis gas, there is a movement to develop the next generation of even more efficient gas turbines ready for carbon dioxide capture-based IGCC power plants.

The heat recovery steam generator captures heat in the hot exhaust from the gas turbines and uses it to generate additional steam that is used to make more power in the steam turbine portion of the combined cycle unit. In most IGCC power plant designs, steam recovered from the gasification process is superheated in the heat recovery steam generator to increase overall efficiency output of the steam turbines. This IGCC combination, which includes a gasification plant, two types of turbine generators (gas and steam), and the heat recovery steam generator, is clean and efficient.

Biomass fuel producers, coal producers, and, to a lesser extent, waste companies are enthusiastic about supplying co-gasification power plant and realize the benefits of co-gasification with alternative fuels. The benefits of a co-gasification technology involving coal and biomass include use of a reliable coal supply with gate-fee waste and biomass which allows greater economies of scale from a larger plant than could be supplied just with waste and biomass. In addition, the technology offers a future option to refineries for hydrogen production and fuel development. In fact, oil refineries and petrochemical plants that have a steady demand for hydrogen can benefit from the installation of gasifiers (Speight, 2011a). The benefits of a co-gasification technology involving coal and biomass include use of a reliable coal supply with gate-fee waste and biomass which allows the economies of scale from a larger plant than could be supplied just with waste and biomass.

4.2 Gasification of Coal with Biomass and Waste

Pyrolysis and gasification of fossil fuels, biomass materials, and wastes have been used for many years to convert organic solids and liquids into useful gaseous, liquid, and cleaner solid fuels (Speight, 2011a; Brar et al., 2012).

4.2.1 Biomass

Biomass includes a wide range of materials, including energy crops such as switch grass and miscanthus, agricultural sources such as corn husks, wood pellets, lumbering and timbering wastes, yard wastes, construction and demolition waste, and bio-solids (treated sewage sludge). Gasification helps recover the energy locked in these materials. Gasification can convert biomass into electricity and products, such as ethanol, methanol, fuels, fertilizers, and chemicals. Thus, in addition to using the traditional feedstocks such as coal and petroleum coke, gasifiers can be designed to utilize biomass, such as yard and crop waste, bio-solids, energy crops, such as switch grass, and waste and residual pulp/paper plant materials as feedstock.

Biomass gasification has been the focus of research in recent years to estimate efficiency and performance of the gasification process using various types of biomass such as sugarcane residue (Gabra et al., 2001), rice hulls (Boateng et al., 1992), pine sawdust (Lv et al., 2004), almond shells (Rapagnà and Latif, 1997; Rapagnà et al., 2000), wheat straw (Ergudenler and Ghaly, 1993), food waste (Ko et al., 2001), and wood biomass (Pakdel and Roy, 1991; Chen et al., 1992; Bhattacharya et al., 1999; Hanaoka et al., 2005). Recently, there has been significant research interest in co-gasification of various biomass and coal mixtures such as Japanese cedar wood and coal, coal and sawdust (Vélez et al., 2009), coal and pine chips (Pan et al., 2000), coal and silver birch wood (Collot et al., 1999), and coal and birch wood (Brage et al., 2000). Co-gasification of coal and biomass has some synergy – the process not only produces a low carbon footprint on the environment, but also improves the H_2/CO ratio in the produced gas which is required for liquid fuel synthesis (Sjöström et al., 1999; Kumabe et al.,

2007). In addition, inorganic matter present in biomass catalyzes the gasification of coal. However, co-gasification processes require custom fittings and optimized processes for the coal and region-specific wood residues.

Biomass usually contains a high percentage of moisture (along with carbohydrates and sugars). The presence of high levels of moisture in the biomass reduces the temperature inside the gasifier, which then reduces the efficiency of the gasifier. Therefore, many biomass gasification technologies require that the biomass be dried to reduce the moisture content prior to feeding into the gasifier.

Like many solid feedstocks, biomass can come in a range of sizes. In many biomass gasification systems, the biomass must be processed to a uniform size or shape to feed into the gasifier at a consistent rate and to ensure that as much of the biomass is gasified as possible. However, beyond the issue of biomass availability, including the seasonal factors associated with many of the biomass feedstocks, another major concern is that more energy is expended in the collection and preparation stages than is generated through processing the biomass although technical hurdles to biomass use remain. In general, there is the general perception in many countries that increased use of biomass feedstocks is driven more by environmental and regulatory factors, if anything, than by free-market forces. Without tax credits or similar incentives, biomass is unlikely to be used as a base-load feedstock and market entry is likely to be as a blend with other feedstocks.

Most biomass gasification systems use air instead of oxygen for the gasification reactions (which is typically used in large-scale industrial and power gasification plants). Gasifiers that use oxygen require an air separation unit to provide the gaseous/liquid oxygen; this is usually not cost-effective at the smaller scales used in biomass gasification plants. Air-blown gasifiers use the oxygen in the air for the gasification reactions.

In general, biomass gasification plants are much smaller than the typical coal or petroleum coke gasification plants used in the power, chemical, fertilizer, and refining industries. As such, they are less expensive to build and have a smaller *facility footprint*. While a large industrial gasification plant may take up 150 acres of land and process 2500 to 15 000 tons per day of feedstock (such as coal or petroleum coke), the smaller biomass plants typically process 25 to 200 tons of feedstock per day and take up less than 10 acres.

While co-gasification of coal and biomass is advantageous from a chemical point of view, some practical problems have been associated with the process on upstream, gasification, and downstream processes. On the upstream side, the particle size of the coal and biomass is required to be uniform for optimum gasification. In addition, moisture content and pretreatment (torrefaction) are very important during upstream processing.

While upstream processing is influential from a material handling point of view, the choice of gasifier operation parameters (temperature, gasifying agent, and catalysts) dictate the product gas composition and quality. Biomass decomposition occurs at a lower temperature than coal and therefore different reactors compatible with the feedstock mixture are required (Brar et al., 2012). Furthermore, feedstock and gasifier type along with operating parameters not only decide product gas composition but also dictate the amount of impurities to be handled downstream. Downstream processes need to be modified if coal is used with biomass in gasification. Heavy metal and impurities such as sulfur and mercury present in coal can make synthesis gas difficult to use and unhealthy for the environment. Also, at high temperature, alkali present in biomass can cause corrosion problems in downstream pipes. An alternative option to downstream gas cleaning would be to process coal to remove mercury and sulfur before feeding it to the gasifier.

However, first and foremost, coal and biomass require drying and size reduction before they can be fed into a gasifier. Size reduction is needed to obtain appropriate particle sizes, while drying is required to achieve a moisture content suitable for gasification operations. In addition, densification of the biomass may be done to make pellets and improve density and material flow in the feeder areas.

It is recommended that biomass moisture content should be less than 15% w/w (in some cases, less than 15% w/w) prior to gasification. High moisture content reduces the temperature achieved in the gasification zone, thus resulting in incomplete gasification. Forest residues or wood has a fiber saturation point at 30 to 31% moisture content (dry basis) (Brar et al., 2012). Compressive and shear strength of the wood increases with decreased moisture content below the fiber saturation point. In such a situation, water is removed from the cell wall which causes shrinkage of the cell wall. The long-chain molecules which make up the cell wall move closer to one another and bind more tightly. A high level of moisture, usually injected in the form of steam in the gasification zone, favors formation of a water − gas shift reaction that increases hydrogen concentration in the resulting gas.

The torrefaction process is a thermal treatment of biomass in the absence of oxygen, usually at 250 to 300°C to drive off moisture, decompose hemicellulose completely, and partially decompose cellulose (Speight, 2011a). Torrefied biomass has reactive and unstable cellulose molecules with broken hydrogen bonds and not only retains 79 to 95% of feedstock energy but also produces a more reactive feedstock with lower atomic hydrogen − carbon and oxygen − carbon ratios than the original biomass. Torrefaction results in higher yields of hydrogen and carbon monoxide in the gasification process.

Finally, the presence of mineral matter in the coal-biomass feedstock is not appropriate for fluidized bed gasification. The low melting point of ash present in woody biomass leads to agglomeration which caused defluidization of the ash and causes sintering, deposition, and corrosion of the gasifier construction metal bed (Vélez et al., 2009). Biomass containing alkali oxides and salts with the propensity to produce yield higher than 5% w/w ash causes clinkering/slagging problems (McKendry, 2002).

Thus, it is imperative to be aware of the melting of biomass ash, its chemistry within the gasification bed (no bed, silica/sand, or calcium bed), and the fate of alkali metals when using fluidized bed gasifiers.

Most small- to medium-sized biomass/waste gasifiers are air-blown and operate at atmospheric pressure and at temperatures in the range 800 to 100°C (1470 to 2190°F). They face very different challenges to large gasification plants − the use of small-scale air separation plant should oxygen gasification be preferred. Pressurized operation, which eases gas cleaning, may not be practical.

Currently, most ethanol in the USA is produced from the fermentation of corn. Vast amounts of corn (and land, water, and fertilizer) are needed to produce the ethanol. As more corn is being used, there is an increasing concern about less corn being available for food. Gasifying biomass, such as corn stalks, husks, and cobs, and other agricultural waste products to produce ethanol and synthetic fuels such as diesel and jet fuel, can help break this energy − food competition.

Thus, the benefits of biomass gasification include: (1) converting what would otherwise be a waste product into high value products, (2) reduced need for landfill space for disposal of solid wastes, (3) decreased methane emissions from landfills, (4) reduced risk of groundwater contamination from landfills, and (5) production of ethanol from non-food sources. Thus, municipalities, as well as the paper and agricultural industries, would be well advised to use gasification to reduce the disposal costs associated with these wastes as well to produce electricity and other valuable products from these waste materials. While still relatively new, biomass gasification shows a great deal of promise. A key advantage of gasification is that it can convert non-food biomass materials, such as corn stalks and wood wastes, to alcohols. Furthermore, biomass gasification does not remove food-based biomass, such as corn, from the economy, unlike typical fermentation processes for making alcohols.

4.2.2 Waste

Municipalities are spending millions of dollars each year on the disposal of solid waste that, in fact, contains valuable unused energy. In addition to the expense of collecting this waste, they must also contend with increasingly limited landfill space, the environmental impact of landfilling, and stringent bans on the use of incinerators. As a result of these challenges, municipalities are increasingly looking to gasification as a solution to help transform this waste into energy.

Waste may be municipal solid waste (MSW), which has had minimal presorting, or refuse-derived fuel (RDF), which has had significant pretreatment, usually mechanical screening and shredding. Other more specific wastes − but excluding hazardous waste − and possibly including petroleum coke, may provide niche opportunities for co-utilization.

The traditional waste-to-energy plant, based on mass-burn combustion on an inclined grate, has a low public acceptability despite the very low emissions achieved over the last decade with modern flue gas clean-up equipment. This has led to difficulty in obtaining planning permission to construct the new waste-to-energy plants that are needed. After much debate, various governments have allowed options for advanced waste conversion technologies (gasification, pyrolysis, and anaerobic digestion), but will only give credit to the proportion of electricity generated from non-fossil waste.

Co-utilization of waste and biomass with coal may provide economies of scale that help achieve the policy objectives identified above at an affordable cost. In some countries, governments propose co-gasification processes as being *well suited for community-sized developments* suggesting that waste should be dealt with in smaller plants serving towns and cities, rather than moved to large, central plants (satisfying the so-called *proximity principal*).

However, at the current time, neither biomass nor wastes are produced, or naturally gathered at sites in quantities sufficient to fuel a modern large and efficient power plant. The disruption, transport issues, fuel use, and public opinion all act against gathering hundreds of megawatts (MWe) at a single location. Biomass or waste-fired power plants are therefore inherently limited in size and hence in efficiency, labor costs per unit electricity produced, and in other economies of scale. The production rates of municipal refuse follow reasonably predictable patterns over time periods of a few years. Recent experience with the very limited current *biomass for energy* harvesting has shown unpredictable variations in harvesting capability with long periods of zero production over large areas during wet weather and the foot and mouth outbreak.

Gasification can convert municipal and other wastes (such as construction and demolition wastes) that cannot be recycled into electric power or other valuable products, such as chemicals, fertilizers, and substitute natural gas. Instead of paying to dispose of these wastes, municipalities are generating income from these waste products, since they are valuable feedstocks for a gasifier. Gasifying municipal and other waste streams reduces the need for landfill space, decreases the generation of methane (a potent greenhouse gas) as the landfill matures through bacterial action, and reduces the potential for groundwater contamination from landfill sites.

The installation of gasifiers by municipalities to dispose of waste and create energy is almost a return to the days when gasification originally became commercial – perhaps déjà vu – and every small town has a gasification plant to produce gas (hence the name *town gas*) for heating and lighting purposes.

However, it is important to add that gasification does not compete with recycling. In fact, gasification complements existing recycling programs by the creation of an added-value product (energy). Many materials, including a wide range of plastics, cannot currently be recycled or recycled any further and are ideal candidates for feedstocks to the gasification process. As the amount of waste generated increases (in line with an increase in the population), recycling rates increase to the point of overburdening the system; gasification will alleviate any potential bottlenecks through the generation of energy.

The situation is very different for coal. This is generally mined or imported and thus large quantities are available from a single source or a number of closely located sources, and supply has been reliable and predictable. However, the economics of new coal-fired power plants of any technology or size have not encouraged any new coal-fired power plant in the gas generation market.

4.2.3 Other Mixed Feedstocks
Combining biomass, refuse, and coal overcomes the potential unreliability of biomass, the potential longer-term changes in refuse, and the size limitation of a power plant using only waste and/or biomass. It also allows benefit from a premium electricity price for electricity from biomass and the gate fee associated with waste. If the power plant is gasification-based, rather than direct combustion, further benefits may be available. These include a premium price for the electricity from waste, the range of technologies available for the gas to electricity part of the process, gas cleaning prior to the main combustion stage instead of after combustion, and public image, which is currently generally better for gasification than for combustion. These considerations have led to the current study of co-gasification of wastes/biomass with coal (Speight, 2008).

For large-scale power generation (>50 MWe), the gasification field is dominated by plant based on the pressurized, oxygen-blown, entrained flow or fixed-bed gasification of fossil fuels. Entrained gasifier operational experience to date has largely been with well-controlled fuel feedstocks with short-term trial work at low co-gasification ratios and with easily handled fuels.

Use of waste materials as co-gasification feedstocks may attract significant disposal credits. Cleaner biomass materials are renewable fuels and may attract premium prices for the electricity generated. Availability of sufficient fuel locally for an economic plant size is often a major issue, as is the reliability of the fuel supply. Use of more predictably available coal alongside these fuels overcomes some of these difficulties and risks. Coal could be regarded as the "flywheel" which keeps the plant running when the fuels producing the better revenue streams are not available in sufficient quantities.

Coal characteristics are very different to younger hydrocarbon fuels such as biomass and wastes. Hydrogen-to-carbon ratios are higher for younger fuels, as is the oxygen content. This means that the reactivity is very different under gasification conditions. Gas cleaning issues can also be very different, with sulfur a major concern for coal gasification but chlorine compounds and tars more important for waste and biomass gasification. There are no current proposals for adjacent gasifiers and gas cleaning systems, one handling biomass or waste and one coal, alongside each other and feeding the same power production equipment. However, there are some advantages to such a design compared with mixing fuels in the same gasifier and gas cleaning system.

Electricity production or combined electricity and heat production remain the most likely area for the application of gasification or co-gasification. The lowest investment cost per unit of electricity generated is the use of the gas in an existing large power station. This has been done in several large utility boilers, often with the gas fired alongside the main fuel. This option allows a comparatively small thermal output of gas to be used with the same efficiency as the main fuel in the boiler as a large, efficient steam turbine can be used. It is anticipated that addition of gas from a biomass or wood gasifier into the natural gas feed to a gas turbine is technically possible but there will be concerns as to the balance of commercial risks to a large power plant and the benefits of using the gas from the gasifier.

The use of fuel cells with gasifiers is frequently discussed but the current cost of fuel cells is such that their use for mainstream electricity generation is uneconomic.

Furthermore, the disposal of municipal and industrial wastes has become an important problem because the traditional means of disposal, landfill, has become environmentally much less acceptable than previously. New, much stricter regulation of these disposal methods will make the economics of waste processing for resource recovery much more favorable. One method of processing waste streams is to convert the energy value of the combustible waste into a fuel. One type of fuel attainable from wastes is a low-heating-value gas, usually $100 - 150$ Btu/scf, which can be used to generate process steam or electricity (Gay et al., 1980). Co-processing such waste with coal is also an option (Speight, 2008).

In summary, coal might be co-gasified with waste or biomass for environmental, technical, or commercial reasons. It allows larger, more efficient plants than those sized for the biomass grown or waste arising within a reasonable transport distance; specific operating costs are likely to be lower and fuel supply security is assured.

Co-gasification technology varies and is usually site specific with high dependence on the feedstock. At the largest scale, the plant may include the well proven fixed-bed and entrained flow gasification processes. At smaller scales, emphasis is placed on technologies which appear closest to commercial operation. Pyrolysis and other advanced thermal conversion processes are included where power generation is practical using the on-site feedstock produced. However, the needs to be addressed are: (1) the core fuel handling and gasification/pyrolysis technologies, (2) the fuel gas clean-up, and (3) the conversion of fuel gas to electric power (Ricketts et al., 2002).

Pyrolysis and gasification of fossil fuels, biomass materials, and wastes have been used for many years to convert organic solids and liquids into useful gaseous, liquid, and cleaner solid fuels.

Gasification generally has a non-negative environmental image whereas combustion has a negative public image. The public view is not always based on objective analysis of emissions and other effects but nevertheless it is still a very important factor. In fact, the technology of co-gasification can result in very clean power plants using a range of fuels but there are considerable economic and environmental challenges.

There is less single-fuel experience with the British Gas Lurgi (BGL) than with entrained gasifiers. However the Lurgi gasifier is better suited to difficult-to-mill feedstocks than are entrained gasifiers and has the most operational experience with fuels of widely differing mechanical properties.

Biomass fuel producers, coal producers, and, to a lesser extent, waste companies are enthusiastic about supplying co-gasification power plant and realize the benefits of co-gasification with, or use on the same site as, "competitor" fuels.

The US has wide experience in gasification technologies, components for fuel handling, general engineering, scientific understanding, and project development. In addition, the technology offers a future option for refineries for hydrogen production and fuel development (Speight, 2011a).

Oil refineries and petrochemical plants provide opportunities for gasifiers when the hydrogen supply is particularly valuable (Speight, 2011b, 2014).

REFERENCES

Agus, H., Sandun, D.F., Sushil A., 2006. Catalytic biomass gasification to produce sustainable hydrogen. Paper No. 066226. Proceedings. American Society of Agricultural and Biological Engineers, St. Joseph, Missouri.

Ancheyta, J., Speight, J.G., 2007. Hydroprocessing of Heavy Oils and Residua. CRC Press, Taylor & Francis Group, Boca Raton, FL.

Anderson, L.L., Tillman, D.A., 1979. Synthetic Fuels from Coal: Overview and Assessment. John Wiley and Sons Inc, New York, p. 33.

Anderson, R.B., 1984. In: Kaliaguine, S., Mahay, A. (Eds.), Catalysis on the Energy Scene. Elsevier, Amsterdam, Netherlands, p. 457.

Argonne. 1990. Environmental Consequences of, and Control Processes for, Energy Technologies. Argonne National Laboratory. Pollution Technology Review No. 181. Noyes Data Corp., Park Ridge, New Jersey. Chapter 6.

Baker, R.T.K., Rodriguez, N.M., 1990. In: Fuel Science and Technology Handbook, Marcel Dekker Inc, New York. Chapter 22.

Batchelder, H.R., 1962. In: McKetta Jr., J.J. (Ed.), Advances in Petroleum Chemistry and Refining, vol. V. Interscience Publishers Inc, New York. Chapter 1.

Bhattacharya, S., Md. Mizanur Rahman Siddique, A.H., Pham, H.-L., 1999. A Study in Wood Gasification on Low Tar Production. Energy 24, 285–296.

Boateng, A.A., Walawender, W.P., Fan, L.T., Chee, C.S., 1992. Fluidized-Bed Steam Gasification of Rice Hull. Bioresour. Technol. 40 (3), 235–239.

Bodle, W.W., Huebler, J., 1981. In: Meyers, R.A. (Ed.), Coal Handbook. Marcel Dekker Inc., New York. Chapter 10.

Brage, C., Yu, Q., Chen, G., Sjöström, K., 2000. Tar Evolution Profiles Obtained from Gasification of Biomass and Coal. Biomass Bioenergy 18 (1), 87−91.

Brar, J.S., Singh, K., Wang, J., Kumar, S., 2012. Cogasification of Coal and Biomass: A Review. Int. J. Forestry Res. 2012 (2012), 1−10.

Cavagnaro, D.M., 1980. Coal Gasification Technology. National Technical Information Service, Springfield, VA.

Chadeesingh, R., 2011. The Fischer − Tropsch Process. In: Speight, J.G. (Ed.), The Biofuels Handbook. The Royal Society of Chemistry, London, UK, pp. 476−517. , Part 3, Chapter 5.

Chen, G., Sjöström, K., Bjornbom, E., 1992. Pyrolysis/Gasification of Wood in a Pressurized fluidized bed reactor. Ind. Eng. Chem. Res. 31 (12), 2764−2768.

Ciferno, J.P., Marano, J.J., 2002. Benchmarking Biomass Gasification Technologies for Fuels, Chemicals and Hydrogen Production. National Energy Technology Laboratory, U.S. Department of Energy, Washington, DC.

Collot, A.G., Zhuo, Y., Dugwell, D.R., Kandiyoti, R., 1999. Co-Pyrolysis and cogasification of coal and biomass in bench-scale fixed-bed and fluidized bed reactors. Fuel 78, 667−679.

Cusumano, J.A., Dalla Betta, R.A., Levy, R.B., 1978. Catalysis in Coal Conversion. Academic Press Inc., New York.

Dry, M.E., 1976. Advances in fischer − tropsch chemistry. Ind. Eng. Chem. Res. Dev. 15 (4), 282−286.

Elton, A., 1958. In: Singer, C., Holmyard, E.J., Hall, A.R., Williams, T.I. (Eds.), A History of Technology, vol. IV. Clarendon Press, Oxford, UK, Chapter 9.

Ergudenler, A., Ghaly, A.E., 1993. Agglomeration of alumina sand in a fluidized bed straw gasifier at elevated temperatures. Bioresour. Technol. 43 (3), 259−268.

Fryer, J.F., Speight, J.G. 1976. Coal Gasification: Selected Abstract and Titles. Information Series No. 74. Alberta Research Council, Edmonton, Canada.

Gabra, M., Pettersson, E., Backman, R., Kjellström, B., 2001. Evaluation of Cyclone Gasifier Performance for Gasification of Sugar Cane Residue − Part 1: Gasification of Bagasse. Biomass Bioenergy 21 (5), 351−369.

Gary, J.G., Handwerk, G.E., Kaiser, M.J., 2007. Petroleum Refining: Technology and Economics, 5th Edition. CRC Press, Taylor & Francis Group, Boca Raton, FL.

Gay, R.L., Barclay, K.M., Grantham, L.F., Yosim, S.J. 1980. Fuel production from solid waste. Symposium on Thermal Conversion of Solid Waste and Biomass. Symposium Series No. 130, American Chemical Society, Washington, DC. Chapter 17, pp. 227−236.

Hanaoka, T., Inoue, S., Uno, S., Ogi, T., Minowa, T., 2005. Effect of woody biomass components on air − steam gasification. Biomass Bioenergy 28 (1), 69−76.

Hotchkiss, R., 2003. Coal gasification technologies. Proc. Inst. Mech. Eng. Part A, 217 (1), 27−33.

Hsu, C.S., Robinson, P.R. (Eds.), 2006. Practical Advances in Petroleum Processing Volume 1 and Volume 2. Springer Science, New York.

Ishi, S., 1982. Coal gasification technology. Energy 15 (7), 40−48.

Kasem, A., 1979. Three Clean Fuels from Coal: Technology and Economics. Marcel Dekker Inc., New York.

Khan, M.R., Patmore, D.J., 1997. Heavy oil upgrading processes. In: Speight, J.G. (Ed.), Petroleum Chemistry and Refining. Taylor & Francis, Washington, DC (Chapter 6).

Ko, M.K., Lee, W.Y., Kim, S.B., Lee, K.W., Chun, H.S., 2001. Gasification of food waste with steam in fluidized bed. Korean J. Chem. Eng. 18 (6), 961−964.

Kumabe, K., Hanaoka, T., Fujimoto, S., Minowa, T., Sakanishi, K., 2007. Cogasification of woody biomass and coal with air and steam. Fuel 86, 684−689.

Lahaye, J., Ehrburger, P. (Eds.), 1991. Fundamental Issues in Control of Carbon Gasification Reactivity. Kluwer Academic Publishers, Dordrecht, Netherlands.

Lee, S., Shah, Y.T., 2013. Biofuels and Bioenergy. CRC Press, Taylor & Francis Group, Boca Raton, FL.

Lee, S., Speight, J.G., Loyalka, S., 2007. Handbook of Alternative Fuel Technologies. CRC-Taylor & Francis Group, Boca Raton, FL.

Long, R., Picioccio, K., Zagoria, A., 2011. Optimizing hydrogen production and use. Petroleum Technol. Q. Autumn, 1−12.

Lv, P.M., Xiong, Z.H., Chang, J., Wu, C.Z., Chen, Y., Zhu, J.X., 2004. An experimental study on biomass air − steam gasification in a fluidized bed. Bioresour. Technol. 95 (1), 95−101.

Mahajan, O.P., Walker Jr., P.L., 1978. In: Karr Jr., C. (Ed.), Analytical Methods for Coal and Coal Products, vol. II. Academic Press Inc., New York, Chapter 32.

Massey, L.G. (Ed.), 1974. Coal Gasification. American Chemical Society, Washington, DC, Advances in Chemistry Series No. 131.

Matsukata, M., Kikuchi, E., Morita, Y., 1992. A new classification of alkali and alkaline earth catalysts for gasification of carbon. Fuel 71, 819−823.

McKendry, P., 2002. Energy production from biomass part 3: gasification technologies. Bioresour. Technol. 83 (1), 55−63.

Mokhatab, S., Poe, W.A., Speight, J.G., 2006. Handbook of Natural Gas Transmission and Processing. Elsevier, Amsterdam, Netherlands.

Nef, J.U., 1957. In: Singer, C., Holmyard, E.J., Hall, A.R., Williams, T.I. (Eds.), A History of Technology, vol. III. Clarendon Press, Oxford, UK, Chapter 3.

Nordstrand, D., Duong, D.N.B., Miller, B.G., 2008. Post-combustion emissions control. In: Miller, B.G., Tillman, D. (Eds.), Combustion Engineering Issues for Solid Fuel Systems. Elsevier, London, UK.

Pakdel, H., Roy, C., 1991. Hydrocarbon content of liquid products and tar from pyrolysis and gasification of wood. Energy Fuels 5, 427−436.

Pan, Y.G., Velo, E., Roca, X., Manyà, J.J., Puigjaner, L., 2000. Fluidized-Bed cogasification of residual biomass/poor coal blends for fuel gas production. Fuel 79, 1317−1326.

Probstein, R.F., Hicks, R.E., 1990. Synthetic Fuels. pH Press, Cambridge, MA, Chapter 4.

Rana, M.S., Sámano, V., Ancheyta, J., Diaz, J.A.I., 2007. A review of recent advances on process technologies for upgrading of heavy oils and residua. Fuel 86, 1216−1231.

Rapagnà, N.J., Latif, A., 1997. Steam gasification of almond shells in a fluidized bed reactor: the influence of temperature and particle size on product yield and distribution. Biomass Bioenergy 12 (4), 281−288.

Rapagnà, N.J., Kiennemann, A., Foscolo, P.U., 2000. Steam-Gasification of biomass in a fluidized-bed of olivine particles. Biomass Bioenergy 19 (3), 187−197.

Ricketts, B., Hotchkiss, R., Livingston, W., Hall, M., 2002. Technology status review of waste/biomass co-gasification with coal. Proceedings Institute of Chemical Engineering Fifth European Gasification Conference. Noordwijk, Netherlands. April 8 − 10.

Sjöström, K., Chen, G., Yu, Q., Brage, C., Rosén, C., 1999. Promoted reactivity of char in cogasification of biomass and coal: synergies in the thermochemical process. Fuel 78, 1189−1194.

Sondreal, E.A., Benson, S.A., Pavlish, J.H., Ralston, N.V.C., 2004. An overview of air quality III: mercury, trace elements, and particulate matter. Fuel Process. Technol. 85, 425−440.

Sondreal, E.A., Benson, S.A., Pavlish, J.H., 2006. Status of research on air quality: mercury, trace elements, and particulate matter. Fuel Process. Technol. 65/66, 5−22.

Speight, J.G., 1990. In: Speight, J.G. (Ed.), Fuel Science and Technology Handbook. Marcel Dekker Inc., New York, Chapter 33.

Speight, J.G., 2007. Natural Gas: A Basic Handbook. GPC Books, Gulf Publishing Company, Houston, TX.

Speight, J.G., 2008. Synthetic Fuels Handbook: Properties, Processes, and Performance. McGraw-Hill, New York.

Speight, J.G. (Ed.), 2011a. The Biofuels Handbook. Royal Society of Chemistry, London, UK.

Speight, J.G., 2011b. The Refinery of the Future. Gulf Professional Publishing, Elsevier, Oxford, UK.

Speight, J.G., 2013a. The Chemistry and Technology of Coal 3rd Edition. CRC Press, Taylor & Francis Group, Boca Raton, FL.

Speight, J.G., 2013b. Coal-Fired Power Generation Handbook. Scrivener Publishing, Salem, MA.

Speight, J.G., 2014. The Chemistry and Technology of Petroleum 5th Edition. CRC Press, Taylor & Francis Group, Boca Raton, FL.

Speight, J.G., Ozum, B., 2002. Petroleum Refining Processes. Marcel Dekker Inc., New York.

Taylor, F.S., Singer, C., 1957. In: Singer, C., Holmyard, E.J., Hall, A.R., Williams, T.I. (Eds.), A History of Technology, vol. II. Clarendon Press, Oxford, UK, Chapter 10.

Van Heek, K.H., Muhlen, H.-J., 1991. In: Lahaye, J., Ehrburger, P. (Eds.), Fundamental Issues in Control of Carbon Gasification Reactivity. Kluwer Academic Publishers Inc, Netherlands, p. 1.

Vélez, F.F., Chejne, F., Valdés, C.F., Emery, E.J., Londoño, C.A., 2009. Cogasification of colombian coal and biomass in fluidized bed: an experimental study. Fuel 88 (3), 424−430.

Wang, Y., Duan, Y., Yang, L., Jiang, Y., Wu, C., Wang, Q., et al., 2008. Comparison of mercury removal characteristic between fabric filter and electrostatic precipitators of coal-fired power plants. J. Fuel Chem. Technol. 36 (1), 23−29.

Yang, H., Xua, Z., Fan, M., Bland, A.E., Judkins, R.R., 2007. Adsorbents for capturing mercury in coal-fired boiler flue gas. J. Hazard. Mater. 146, 1−11.

CHAPTER 5

The Fischer – Tropsch Process

1 INTRODUCTION

Fischer – Tropsch synthesis is a catalyzed chemical reaction in which synthesis gas (syngas), a mixture of carbon monoxide (CO) and hydrogen (H_2), is converted into gaseous, liquid, and solid hydrocarbons and an appreciable amount of oxygenates (Chadeesingh, 2011). This process is a highly promising, developing option for environmentally sound production of chemicals and fuels from biomass, coal, and natural gas. In view of large coal and natural gas reserves, dwindling petroleum reserves, and significant, projected increases in demand for liquid fuels, it is expected to play an ever increasing role in the coming decades. Fischer – Tropsch synthesis can be based on several synthesis gas feedstocks including those from coal gasification, natural gas, and biomass.

Thus, as crude oil production decreases and its price increases, the Fischer – Tropsch (F-T) technology which enables the production of synthetic hydrocarbons from coal or natural gas feedstocks is becoming an increasingly attractive technology in the energy mix. In fact, coupled with this is the fact that Fischer – Tropsch products are ultra-clean fuels in that they contain no aromatics, sulfur, or nitrogen compounds. In essence, compared to petroleum derived gasoline and diesel, the analogous fraction derived from the Fischer – Tropsch process will burn to produce considerably less polyaromatic hydrocarbons (PAHs), and no sulfur oxides (SOx) or nitrogen oxides (NOx).

With the intensification of global pressures to reduce greenhouse gas emissions, legislative frameworks in Europe and the USA have already been put in place to force producers of liquid transportation fuels to comply with stricter emission standards. The impact of such legislation is that dilution of petroleum derived fuels with the cleaner Fischer – Tropsch derived hydrocarbons is becoming an increasingly important way to achieve environmental compliance. It is thus not surprising that Fischer – Tropsch technology now occupies a visible place in the energy mix required for sustainable global development.

The history of Fischer – Tropsch technology dates back to over a century and since the turn of the 21st Century there has been significantly renewed interest in Fischer – Tropsch technology. In great part this renaissance has been due to the exploitation of cheaper remote gas (*stranded gas*), which has the effect of making the economics of Fischer – Tropsch projects increasingly attractive.

A Fischer – Tropsch plant incorporates three major process blocks: (1) production of synthesis gas, i.e. a mixture of carbon monoxide and hydrogen (*steam reforming*), (2) conversion of synthesis gas to aliphatic hydrocarbons and water (*Fischer – Tropsch synthesis process*), and (3) hydrocracking of the longer chain, waxy synthetic hydrocarbons to fuel grade fractions. Of these three steps, the production of synthesis gas is the most energy intensive as well as expensive and this step can account for as much as 50 to 75% of capital costs. While extensive research has been undertaken with the aim of finding process improvements in all three steps, this step has received the greatest attention. In great measure, this has significantly impacted on allowing plants the ability to attain a scale almost inconceivable until recent times. The technological advances relating to this step are thus discussed in addition to the actual Fischer – Tropsch synthesis process. In summary, the Fischer – Tropsch process yields a wide range of hydrocarbons, from low-boiling gases to high-boiling wax.

This chapter discusses the various aspects of the Fischer – Tropsch process and how research and development is ongoing with successes being measured by the demonstration and commercialization of technologies such as a permeable membrane for the generation of high-purity hydrogen, which in itself can be used to adjust the H_2/CO ratio of the synthesis gas produced.

2 PRODUCTION OF SYNTHESIS GAS

The process for producing synthesis gas can be described as comprising three components: (1) synthesis gas generation, (2) waste heat recovery, and (3) gas processing (Chapters 1 and 3). Within each of the above three listed systems are several options. For example, synthesis gas can be generated to yield a range of compositions ranging from high-purity hydrogen to high-purity carbon monoxide. Two major routes can be utilized for high-purity gas production: (1) pressure swing adsorption and (2) utilization of a cold trap or reactor, where separation is achieved by distillation at low temperatures. In fact, both processes can also be used in combination as well. Unfortunately, both processes require high capital expenditure. However, to address these concerns, research and development is ongoing and successes can be measured by the demonstration and commercialization of technologies such as a permeable membrane for the generation of high-purity hydrogen.

2.1 Steam Reforming

Steam reforming (sometimes referred to as *steam methane reforming*, SMR) is carried out by passing a preheated mixture comprising essentially methane and steam through catalyst filled tubes. Since the reaction is endothermic, heat must be provided in order to effect the conversion. This is achieved by the use of burners located adjacent to the tubes. The products of the process are a mixture of hydrogen, carbon monoxide, and carbon dioxide. Recovery of the heat from the combustion products can be implemented in order to improve the efficiency of the overall process.

To maximize the conversion of the methane feed, both a primary and secondary reformer are generally utilized. A *primary reformer* — in which the hydrocarbon feedstock is partially reacted with steam over a nickel – alumina catalyst to produce a synthesis gas with an H_2/CO ratio of approximately 3:1 — is used to effect 90 to 92% conversion of methane. This is done in a fired tube furnace at 900°C (1650°F) at a pressure of up to 450 psi. The unconverted methane is reacted with oxygen at the top of a *secondary autothermal reformer* containing nickel catalyst in the lower region of the vessel.

Two water – gas shift (WGS) reactors are used downstream of the secondary reformer to adjust the H_2/CO ratio, depending on the end use of the steam reformed products. The first of these two reactors utilizes an iron-based catalyst which is heated to approximately 400°C (750°F). The second WGS reactor operates at 200°C (390°F) and is charged with a copper-based catalyst.

The deposition of carbon can be an acute problem with the use of nickel-based catalysts in the primary reformer (Rostrup-Neilsen, 1984; Alstrup, 1988; Rostrup-Neilsen, 1993) and steps must be taken to prevent carbon deposition on the catalyst. A successful technique is to use a steam/carbon ratio in the feedstock gas that does not allow the formation of carbon – however, this method results in lowering the efficiency of the process. Another approach is to utilize sulfur passivation, which led to the development of the SPARG process (Rostrup-Neilsen, 1984; Udengaard et al., 1992). This technique utilizes the principle that the reaction leading to the deposition of carbon requires a larger number of adjacent surface Ni atoms than does steam reforming. When a fraction of the surface atoms are covered by sulfur, the deposition of carbon is thus more greatly inhibited than steam reforming reactions. A third approach is to use Group VIII metals (such as platinum) that do not form carbides but, due to the high cost of such metals, they do not compare to the economics associated with the use of nickel.

A major challenge in steam reforming development is the energy intensive nature of the process due to the high endothermic character of the reactions. The trend in development thus is one which seeks higher energy efficiency. Improvements in catalysts and metallurgy require adaption to lower steam/carbon ratios and higher heat flux.

2.2 Autothermal Reforming

Autothermal reforming (ATR) was developed in the 1950s and is used in commercial applications to provide synthesis gas for ammonia and methanol synthesis (Chukwu, 2002; Krumpelt et al., 2002). In the case of ammonia production, where high H_2/CO ratios are needed, the autothermal reforming process is operated at high steam/carbon ratios. In the case of methanol synthesis, the required H_2/CO ratio is provided by manipulating the carbon dioxide recycle. In fact, development and optimization of this technology has led to cost-effective operation at very low steam/carbon feed ratios to produce carbon monoxide-rich synthesis gas, for example, that which is preferred in Fischer – Tropsch synthesis.

In the autothermal reforming process, the organic feedstock (e.g., natural gas) and steam (and sometimes carbon dioxide) are mixed directly with oxygen and air in the reformer. The reformer itself comprises a refractory lined vessel which contains the catalyst, together with an injector located at the top of the vessel. Partial oxidation reactions occur in a region of the reactor referred to as the combustion zone. It is the mixture from this zone which then flows through a catalyst bed where the actual reforming reactions occur. Heat generated in the combustion zone from partial oxidation reactions is utilized in the reforming zone, so that in the ideal case, it is possible that the auto-thermal reforming process can exhibit heat balance.

When the autothermal reformer uses carbon dioxide, the H_2:CO ratio produced is 1:1; when the autothermal reformer uses steam, the H_2:CO ratio produced is 2.5:1. The reactions can be described in the following equations, using CO_2:

$$2CH_4 + O_2 + CO_2 \rightarrow 3H_2 + 3CO + H_2O + heat$$

Using steam:

$$4CH_4 + O_2 + 2H_2O \rightarrow 10H_2 + 4CO$$

The reactor itself consists of three zones: (1) the burner − here the feed streams are mixed in a turbulent diffusion flame, (2) the combustion zone − where partial oxidation reactions produce a mixture of carbon monoxide and hydrogen, and (3) the catalytic zone − where the gases leaving the combustion zone reaches thermodynamic equilibrium.

The following are the advantages of using the autothermal reforming process: (1) compact in design, hence less associated footprint, (2) low investment, (3) economy of scale, (4) flexible operation − short startup periods and fast load changes, and (5) soot-free operation.

2.3 Combined Reforming

Combined reforming incorporates the combination of both steam reforming and autothermal reforming. In such a configuration, the hydrocarbon (e.g., natural gas) is first only partially converted, under mild conditions, to synthesis gas in a relatively small steam reformer. The off-gas stream from the steam reformer is then sent to an oxygen-fired secondary reactor, the autothermal reforming reactor. Here, the unreacted methane is converted to synthesis gas by partial oxidation followed by steam reforming.

Another configuration requires the hydrocarbon feed to be split into two streams which are then fed in parallel to the steam reforming and autothermal reactors.

2.4 Thermal Partial Oxidation

Partial oxidation (POX or P_{OX}) reactions occur when a sub-stoichiometric fuel-air mixture is partially combusted in a reformer.

The general reaction equation without catalyst (*thermal partial oxidation*, TPOX) is of the form:

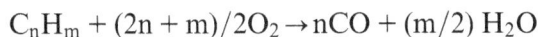

$$C_nH_m + (2n + m)/2O_2 \rightarrow nCO + (m/2) H_2O$$

A possible reaction equation (coal):

$$CH_{feedstock} + O_2 \rightarrow CO + 6H_2$$

A TPOX reactor is similar to the autothermal reactor – the main difference being no catalyst is used. The feedstock, which may include steam, is mixed directly with oxygen by an injector which is located near the top of the reaction vessel. Both partial oxidation reactions as well as reforming reactions occur in the combustion zone below the burner.

The principal advantage of the partial oxidation process is the ability of the process to accept a wide variety of feedstocks, which can comprise very high molecular weight organics, for example petroleum coke (Gunardson and Abrardo, 1999). Additionally, since emission of NOx and SOx are minimal, the technology can be considered environmentally benign.

On the other hand, very high temperatures, approximately 1300°C (2370°F), are required to achieve near complete reaction. This necessitates the consumption of some of the hydrogen and a greater than stoichiometric consumption of oxygen, i.e., oxygen rich conditions. Capital costs are high on account of the need to remove soot and acid gases from the synthesis gas. Operating expenses are also high due to the need for oxygen at high pressure.

2.5 Catalytic Partial Oxidation

A possible means of improving the efficiency of synthesis gas production is via *catalytic partial oxidation* (CPOX, CP_{OX}) technology. The concept has several advantages over steam reforming, especially the higher energy efficiency. The reaction is in fact, not endothermic as is the case with steam reforming, but rather slightly exothermic. Further, an H_2/CO ratio of close to 2.0, i.e., the ideal ratio for the Fischer – Tropsch and methanol synthesis, is produced by this technology. The process can occur by either of two routes: (1) direct catalytic partial oxidation or (2) indirect catalytic partial oxidation.

2.5.1 Direct Catalytic Partial Oxidation

Direct catalytic partial oxidation involves a mechanism in which only surface reactions on the catalyst are operative and produces synthesis:

$$2CH_4 + O_2 \rightarrow 2CO + 2H_2$$

The *direct* mechanism is likely to occur at short contact times and higher than equilibrium values are obtained with high flow rates through fixed bed reactors (Choudhary et al., 1993; Lapszewicz and Jiang, 1992).

2.5.2 Indirect Catalytic Partial Oxidation

The indirect route comprises total combustion of methane to carbon dioxide and water (which is not partial oxidation but complete oxidation), followed by steam reforming and the water − gas shift reaction. Here, equilibrium conversions can be greater than 90% at ambient pressure. However, in order for an industrial process for this technology to be economically viable, an operating pressure in excess of 300 psi would be required. Unfortunately, under such pressures, equilibrium conversions are lower. Further, an operational problem arises on account of the highly exothermic combustion step, which makes for problematic temperature control of the process and the possibility of temperature runaways.

2.6 Membrane Reactors

An innovative technology for combining air separation and natural gas reforming processes involves the use of membrane reactors (Dyer and Chen, 1999; Nataraj et al., 2000). The technology is referred to as oxygen transport membranes (OTM) and should combine five unit operations currently in use: (1) oxygen separation, (2) oxygen compression, (3) partial oxidation, (4) steam methane reforming, and (5) heat exchange. This technology incorporates the use of catalytic components with the membrane to accelerate the reforming reactions.

This technology can be utilized to generate synthesis gas from several feedstocks, including natural gas, associated gas (from crude oil production), light hydrocarbon gases from refineries, and medium weight hydrocarbon fractions such as naphtha. The first stage comprises conventional steam reforming with partial conversion to synthesis gas. This is followed by complete conversion in an ion transport ceramic membrane (ITM) reactor. This combination solves the problem associated with steam reforming for feedstocks with hydrocarbons heavier than methane, since C^{2+} hydrocarbons tend to crack and degrade both the catalyst and membrane.

By shifting the equilibrium in the steam reforming process through removal of hydrogen from the reaction zone, membrane reactors can also be used to increase the equilibrium-limited methane conversion. Using a palladium – silver (Pd – Ag) alloy membrane reactor, methane conversion can reach in excess of 90% v/v (Shu et al., 1995).

3 PRODUCTION OF PURE CARBON MONOXIDE AND HYDROGEN

Purities in excess of 99.5% v/v of either the hydrogen or carbon monoxide produced from synthesis gas can be achieved if desired. Four of the major process technologies available are:

1. *Cryogenics + Methanation*, which utilizes a cryogenic process (occurring in a cold box) whereby carbon monoxide is liquefied in a number of steps until hydrogen with a purity of approximately 98% v/v is produced. The condensed carbon monoxide product, which would contain methane, is then distilled to produce essentially pure carbon monoxide and a mixture of carbon monoxide and methane. The latter stream can be used as fuel. The hydrogen stream from the cold box is taken to a shift converter where the remaining carbon monoxide is converted to carbon dioxide and hydrogen. The carbon dioxide is then removed and any further carbon monoxide or carbon dioxide can be removed by methanation. The resulting hydrogen stream can have a purity on the order of 99.7% v/v.
2. *Cryogenics plus pressure swing adsorption (PSA)*, which utilizes the similar sequential liquefaction of carbon monoxide in a cold box until hydrogen of approximately 98% v/v purity is achieved. Again, the carbon monoxide stream can be further distilled to remove methane until it is essentially pure. The hydrogen stream is then allowed to go through multiple pressure swing adsorption cycles until hydrogen with a purity as high as 99.999% v/v is produced.
3. *Methane-wash cryogenic process*, which involves a process in which liquid carbon monoxide is absorbed into a liquid methane stream so that the hydrogen stream produced contains only ppm levels of carbon

monoxide but approximately 5 to 8% v/v methane. Hence the purity of the hydrogen stream is on the order of 95% v/v. However, the liquid carbon monoxide/methane stream can be distilled to produce an essentially pure carbon monoxide stream and a carbon monoxide/methane stream which can be used as fuel.

4. *COsorb process*, which utilizes copper ions (cuprous aluminum chloride, $CuAlCl_4$) in toluene to form a chemical complex with carbon monoxide, and in effect separates it from the hydrogen, nitrogen, carbon dioxide, and methane. This process can capture about 96% v/v of the carbon monoxide to produce a hydrogen stream with a purity in excess of 99% v/v. The disadvantage of this process is that water, hydrogen sulfide, and other trace chemicals which can poison the copper catalyst must be removed prior to entry of the gas stream into the absorption reactor. Furthermore, the purity of the hydrogen stream is relatively low — on the order of 97% v/v. However, while the efficiency of cryogenic separation decreases with low carbon monoxide content of the feed, the COsorb process is able to process gases with low carbon monoxide content more efficiently.

4 FISCHER – TROPSCH CHEMISTRY

As a simplification, the Fischer – Tropsch synthesis is essentially a polymerization reaction in which carbon bonds are formed from carbon atoms derived from carbon monoxide, under the influence of hydrogen in the presence of a metal catalyst. The reaction leads to a range of products which depend on the reaction conditions and catalysts employed.

Despite many years of extensive study on Fischer – Tropsch synthesis, there remains a lack of deep insight into the fundamental reaction behavior of Fischer – Tropsch synthesis, such as the main reasons for causing the large changes in the reaction performance in the early stage and the reason why Fischer – Tropsch synthesis has a very unique product distribution as well as the role of the products (hydrocarbons and water) in the performance of Fischer – Tropsch synthesis. Such problems have not been fully resolved and further work is necessary to explain these complex phenomena (Overett et al., 2000; Dry, 2004).

4.1 General Aspects

For the most part, the Fischer – Tropsch reaction can be described as the synthesis of hydrocarbons via the hydrogenation of carbon monoxide using transition metal catalysts, such as iron, cobalt, ruthenium, and nickel. Simply, the reactions can be regarded as a carbon chain building process where methylene (CH_2) groups are attached sequentially in a carbon chain (Mukoma et al., 2006):

$$nCO + [n + m/2]H_2 \rightarrow C_nH_m + n\,H_2O \quad -\Delta H$$

For example:

$$CO + 2H_2 \rightarrow -CH_2- \; + H_2O \quad \Delta H = -165 \text{ kJ/mol}$$

Several other reactions also occur and the mechanism is not fully understood (Table 5.1) (Spath and Dayton, 2003).

As a *general rule of thumb*, the reactions which produce water and carbon dioxide tend to be more exothermic on account of the very high heat of formation of these species. Some of the reactions proposed are as follows (Rauch, 2001):

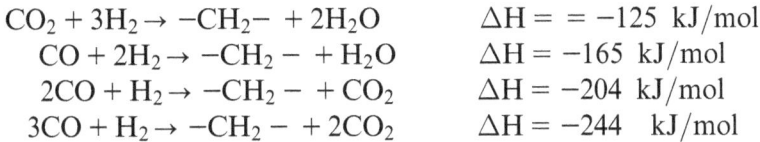

$$\begin{aligned}
CO_2 + 3H_2 &\rightarrow -CH_2- \; + 2H_2O & \Delta H &= = -125 \text{ kJ/mol} \\
CO + 2H_2 &\rightarrow -CH_2 - \; + H_2O & \Delta H &= -165 \text{ kJ/mol} \\
2CO + H_2 &\rightarrow -CH_2 - \; + CO_2 & \Delta H &= -204 \text{ kJ/mol} \\
3CO + H_2 &\rightarrow -CH_2 - \; + 2CO_2 & \Delta H &= -244 \text{ kJ/mol}
\end{aligned}$$

There is also the water – gas shift reaction:

$$CO + H_2O \rightarrow H_2 + CO_2 \quad \Delta H = = -39 \text{ kJ/mol}$$

Table 5.1 Reactions Occurring During the Fischer – Tropsch Synthesis	
Main Reactions	
1. Paraffins	$(2n + 1) H_2 + nCO \rightarrow C_nH_{2n+2} + nH_2O$
2. Olefins	$2nH_2 + nCO \rightarrow C_nH_{2n} + nH_2O$
3. Water gas shift reaction	$CO + H_2O \rightleftarrows CO_2 + H_2$
Side Reactions	
4. Alcohols	$2nH_2 + nCO \rightarrow C_nH_{2n+2}O + (n-1)\,H_2O$
5. Boudouard reaction	$2CO \rightarrow C + CO_2$
Catalyst Modifications	
6. Catalyst oxidation/reduction	a. $M_xO_y + yH_2 \rightleftarrows yH_2O + xM$
	b. $M_xO_y + yCO \rightleftarrows yCO_2 + xM$
7. Bulk carbide formation	$yC + xM \rightleftarrows M_xC_y$

Due to the very high exothermic nature of the Fischer-Tropsch reactions an important issue is the need to avoid an increase in temperature (Fürnsinn et al., 2005). The need for cooling is thus of critical importance in order to: (1) maintain stable reaction conditions, (2) avoid the tendency to produce lighter hydrocarbons, and (3) prevent catalyst sintering and hence reduction in activity. Furthermore, since the total heat of reaction is in the order of approximately 25% of the heat of combustion of the synthesis gas (i.e., reactants in the process), a theoretical limit on the maximum efficiency of the Fischer − Tropsch process is imposed (Rauch, 2001).

Generally, four types of catalysts are used to catalyze the Fischer − Tropsch reaction and they are nickel, cobalt, iron, and ruthenium, which each have a different ability to favor certain reactions although the reaction conditions also have a strong effect on them. Nickel catalysts are highly selective to methane compared to the cobalt, iron, and ruthenium catalysts. Under typical conditions (e.g., 180 to 270°C, 355 to 520°F, at a hydrogen/carbon monoxide ratio of 1 to 2) the cobalt, iron, and ruthenium catalysts promote the formation of paraffins and olefins reactions. The selectivity of these catalysts for olefin formation is in the sequence: $Ru > Fe > Co > Ni$ and iron catalysts have the highest alcohol selectivity for alcohol production. Of these four metals, only iron catalyzes the water − gas shift reaction under typical reaction conditions, thereby enabling operation at a lower hydrogen/carbon monoxide ratio.

4.2 Low- and High-Temperature Fischer − Tropsch Processes

In practice, there are two Fischer − Tropsch process options, each dictated by the type of catalyst used and the temperature ranges utilized. If the catalyst used is iron-based, a temperature range of 300 to 350°C (570 to 650°F) is used and constitutes the high-temperature process (HTFT). On the other hand, if the catalyst selected is cobalt-based, then the temperature range required is on the order of 200 to 240°C (390 to 465°F) (Iglesia, 1997; Dry, 2002).

In principle, other catalysts can also be used in the Fischer − Tropsch synthesis, especially those with rubidium, Ru, or, nickel, Ni, active sites. In practice, however, on account of the low availability of rubidium and consequently its cost, this type of catalyst has not managed to find a place in commercial scale applications even though its activity is sufficient for a successful Fischer − Tropsch process. Nickel-based catalysts on the other hand, while having high enough activities for commercial scale application, suffer from the fact that they tend to produce too much methane. Also, at high pressures (approximately 450 psi) its performance is considered poor due to the tendency for the production of volatile oxygenates.

In practice, the two major catalysts used in industry remain those which are either iron-based or cobalt-based. From the commercial standpoint, iron is cheap, and while cobalt may be considerably more expensive, this is offset by the fact that it offers greater activity as well as longer life, negating the need for frequent planned plant shutdowns to enable change of catalysts (Dry, 2001).

4.3 Reactor Design

Fixed bed or slurry bed reactors are the two conventional reactor types currently used for Fischer – Tropsch processes. Fixed bed reactors are comparable to the shell and tube heat exchangers that are common in the process industries. In these the catalyst, in the form of cylindrical pellets, is contained within 1- to 2-inch tubes that are oriented vertically within a large vessel. In the latest fixed bed Fischer – Tropsch reactors, these tubes are 1 inch in diameter and approximately 40 feet high.

The greatest challenge with conventional fixed bed Fischer – Tropsch reactors lies in controlling the temperature within the reactor tubes. Since the Fischer – Tropsch synthesis is exothermic and strongly affected by temperature, a hot spot can develop within the center of the tube, resulting in a substantial drop in the catalyst performance. As a result, the performance of fixed bed reactors is limited by heat transfer.

In slurry bed reactors, the Fischer – Tropsch catalysts take the form of small particulates (approximately 50 μm diameter) that are suspended in the liquid wax produced by the reaction. The heat generated by the reaction is removed by coolant tubes that run throughout the reactor. The liquid slurry is quite efficient at heat removal; however, the liquid film surrounding the catalyst blocks the reactants (hydrogen and carbon monoxide) from quickly reaching the catalytic sites. This problem with mass transfer limits the performance and productivity of the slurry bed reactors. Additionally, these reactors can be more difficult to operate than fixed bed reactors. A particular problem associated with slurry bed reactors is the catalyst attrition caused when the catalyst particles collide against each other and the reactor walls.

Micro-channel reactors are based on the use of micro-channel process technology, a developing field of chemical processing that exploits rapid reaction rates by minimizing heat and mass transport limitations. This is achieved by reducing dimensions of the reactor systems. In micro-channel reactors the key process steps are carried out in parallel arrays of micro-channels, each with typical dimensions less than 0.2 inch. This modular structure offers many advantages when it comes to reducing the size and cost of the chemical processing hardware. In addition, the modular structure means that maintenance and catalyst replacement can be carried out by replacing individual modules, rather than requiring the prolonged shutdown of the entire system (Atkinson and McDaniel, 2009).

The issues which govern the design of reactors best suited to large scale production of Fischer − Tropsch products are: (1) heat removal arising out of the exothermic reactions and (2) temperature control. These are important essentially to enable longer catalyst lifetimes and in obtaining optimal product selectivity.

The overall product selectivity is controlled by the process operating temperature, reactant partial pressures, and, in the case of iron catalysts, the level of the alkali promoter. In iron catalysts, the higher the alkali content the greater the shift to higher-carbon-number products. For both cobalt and iron catalysts targeting wax production at lower temperatures, the hydrogen/carbon monoxide ratio is a key factor. The operating pressure does not affect product selectivity for iron catalysts, whereas for cobalt, there is a shift toward higher molecular mass products as the total pressure increases (Dry, 1982). It has also been reported that for the Fischer − ropsch process operating at high temperature, especially using an iron catalyst, the situation is more complicated. Apart from the total pressure, product selectivity also is dependent on the partial pressures of hydrogen, carbon monoxide, water, and carbon dioxide (Dry, 2004).

There are currently four major reactor designs that have been developed in order to maximize the efficiency of heat removal and enable optimal temperature control (Dry, 2001): (1) the multi-tubular fixed bed reactor, (2) the fixed slurry bed reactor, (3) the circulating fluidized bed (Synthol) reactor, and (4) the fixed-fluidized bed (Sasol) reactor.

In fact, the effect of designing a Fischer – Tropsch process to produce specific chain length products has been studied using simplified flow sheet models (Mukoma et al., 2006). Designing a Fischer – Tropsch process with recycling and reforming of the lighter gases is very beneficial, especially if the reaction products are described by low to middle chain growth values, because higher carbon efficiency values are obtained and the amount of overall CO conversion is enhanced.

Thus for a fixed production rate of liquid fuels at 100% conversion of carbon monoxide, the carbon efficiency for the process with a recycle stream is higher than that of the once-through process. However, when the aim of the process is to maximize diesel production by hydrocracking the waxes, an optimal chain growth is necessary to reduce the cost of hydrocracking very heavy waxes. The incorporation of wax hydrocracking into the two processes reduces the carbon efficiency at all chain growth values, thereby making it uneconomical to produce very-long-chain hydrocarbons.

In addition, the design and operating conditions for the two processes will be different. From the temperature point of view, the once-through process will be enhanced if it is a low-temperature Fischer – Tropsch process, whereas the recycle process will be enhanced if it is a high-temperature Fischer – Tropsch process because the mean chain lengths and, therefore, the hydrocarbon distribution of Fischer – Tropsch synthesis changes with temperature (Mukoma et al., 2006). At higher temperatures, the mean chain length is smaller, and more methane and lighter hydrocarbons and less diesel and waxes are formed. A cobalt-based catalyst that is used as a low-temperature Fischer-Tropsch catalyst will be suited for the once-through process, to maximize wax production, whereas the iron catalyst will be ideal for the high-temperature process with a recycle stream.

4.4 Product Distribution

When the products desired are the shorter carbon chain lengths, e.g., the light petroleum or gasoline fractions, the longer chain groups can be cracked accordingly. It would thus appear that Fischer – Tropsch synthesis conditions which result in product distributions that provide longer carbon chains is more amenable a greater flexibility in choosing saleable fractions of choice.

While there is considerable support for the stepwise oligomerization mechanism, it does not provide a suitable explanation of large amounts of oxygenated compounds (alcohols, aldehydes, ketones, and carboxylic acids) which appear to be primary products of the Fischer – Tropsch process. It is evident that the mechanism suffers from significant deficiencies. As a result, the issue of an all encompassing Fischer–Tropsch mechanism remains elusive and continues to remain a controversial issue in the literature (Anderson, 1956; Dry, 1990, 1996; Chadeesingh, 2011).

4.5 Upgrading Fischer – Tropsch Liquids

Conventional refinery processes can be used for upgrading of Fischer – Tropsch liquid and wax products. Low-temperature Fischer – Tropsch synthesis is employed in the production of waxes, which are converted into naphtha or diesel oil after the hydroprocessing. High-temperature Fischer – Tropsch synthesis is employed in the production of gasoline and alpha-olefins. Fuels produced with Fischer – Tropsch synthesis are of a high quality due to a very low aromaticity and zero sulfur content. The product stream consists of various fuel types: liquefied petroleum gas (LPG), gasoline, diesel fuel, jet fuel. The definitions and conventions for the composition and names of the different fuel types are obtained from crude oil refinery.

Hydroprocessing is employed in the treatment of wax produced by Fischer – Tropsch synthesis – the wax is composed of linear paraffins as well as small quantities of olefins and oxygenates. The hydrogenation of the olefins and the oxygenated compounds, besides hydrocracking of wax, can be conducted in conditions that are not very severe, with the production of naphtha and diesel oil.

Upgrading of the Fischer – Tropsch products from the fixed bed and slurry bed reactors becomes necessary for the conversion of heavier waxes into usable and lighter products in the diesel range. Product upgrading is an optional stage added to the Fischer – Tropsch synthesis of gas-to-liquid (GTL) to increase the conversion and premium of the end product; it might not be necessary when a fluid bed reactor is used for paraffin synthesis (Chukwu, 2002). For the separate production of Fischer – Tropsch fuels from natural gas, the synthesis gas is then catalytically combined over Fischer – Tropsch catalysts to produce hydrocarbon fuels in a slurry-phase Fischer − Tropsch reactor. The unconverted synthesis gas is then recycled to the synthesis unit to complete the conversion to hydrocarbon fuels.

REFERENCES

Alstrup, I., 1988. New model explaining carbon filament growth on Nickel, Iron, and Ni-Cu alloy catalysts. J. Catal. 109 (2), 241−251.

Anderson, R.B., 1956. Catalysts for the fischer − tropsch synthesis. In: Emmet, P.H. (Ed.), Catalysis Volume IV, Hydrocarbon Synthesis, Hydrogenation and Cyclization. Reinhold, New York (Chapter 2).

Atkinson, D., McDaniel, J., 2009. Fischer − tropsch: micro-channel reactors lead to mega biofuel benefits. Chem. Eng. World January, 28−36.

Chadeesingh, R., 2011. The fischer − tropsch process. The Biofuels Handbook. The Royal Society of Chemistry, London, UK (Part 3, Chapter 5).

Choudhary, V.R., Rajput, A.M., Prabhakar, B., 1993. Non-equilibrium Oxidative Conversion of Methane to CO and H_2 with High Selectivity and Productivity over Ni/Al_2O_3 at Low Temperatures. J. Catal. 139, 326−328.

Chukwu, G.A., 2002. Study of Transportation of GTL Products from Alaskan North Slope (ANS) To Markets. Fischer − Tropsch Archive, Report Archive, United States Department of Energy, Washington, DC. September.

Dry, M.E., 1982. Catalytic Aspects of Industrial Fischer − Tropsch Synthesis. J. Mol. Catal. 17, 133−144.

Dry, M.E., 1990. The fischer − tropsch process: commercial aspects. Catal. Today 6, 183−206.

Dry, M.E., 1996. Practical and theoretical aspects of the catalytic fischer − tropsch process. Appl. Catal. A: Gen. 138, 319−344.

Dry, M.E., 2001. High Quality Diesel via the FT Process: A Review. J. Chem. Technol. Biotechnol. 77, 43−50.

Dry, M.E., 2002. The fischer − tropsch process: 1950 − 2000. Catal. Today 71, 227−241.

Dry, M.E., 2004. Present and future applications of the fischer − tropsch process. Appl. Catal. A 276, 1−3.

Dyer, P.N., Chen, C.M., 1999. Engineering development of ceramic membrane reactor systems for converting natural gas to hydrogen and synthesis gas for transportation fuels. Proceedings Energy Products for the 21st Century Conference, Sept 22.

Fürnsinn, S., Ripfel, K., Rauch, R., Hofbauer, H., 2005. Diesel aus Holz − Die FS Synthese als zukunftsweisende Technologie zur Gewinnung flüssiger Brennstoffe aus Biomasse, 4. Internationale Energiewirtschaftstagung an der TU Wien.

Gunardson, H.H., Abrardo, J.M., 1999. CO-rich synthesis gas. Hydrocarbon Processing April, 87−93.

Iglesia, E., 1997. Fischer − tropsch synthesis on cobalt catalysts: structural requirements and reaction pathways. Stud. Surf. Sci. Catal. 107, 153.

Krumpelt, M., Krause, T., Kopasz, J., Carter, D., Ahmed, S., 2002. Catalytic autothermal reforming of hydrocarbon fuels for fuel cells. Proceedings American Institute of Chemical Engineering, Spring National Meeting, Fuel Processing Session II, New Orleans, Louisiana. March 10 − 14.

Lapszewicz, J.A., Jiang, X., 1992. Mechanism of partial oxidation of methane to synthesis Gas. Preprints ACS Div. Petr. Chem. 37, 252.

Mukoma, P., Hildebrandt, D., Glasser, D., 2006. A process synthesis approach to investigate the effect of the probability of chain growth on the efficiency of fischer − tropsch synthesis. Ind. Eng. Chem. Res. 45, 5928−5935.

Nataraj, S., Moore, R.B., Russek, S.L., 2000. Production of Synthesis Gas by Mixed Conducting Membranes. United States Patent 6,048,472. April 11.

Overett, M.J., Hill, R., Moss, J., 2000. Organometallic chemistry and surface science: mechanistic models for the fischer − tropsch synthesis. Coord. Chem. Rev. 206-207, 581−605.

Rauch, R., 2001. Biomass Gasification to produce Synthesis Gas for Fuel Cells, Liquid Fuels and Chemicals, IEA Bioenergy Agreement, Task 33: Thermal Gasification of Biomass.

Rostrup-Neilsen, J.R., 1984. Sulfur-passivated nickel catalysts for carbon-free steam reforming of methane. J. Catal. 85, 31−43.

Rostrup-Neilsen, J.R., 1993. Production of synthesis gas. Catal. Today 19, 305−324.

Shu, J., Grandjean, B.P.A., Kaliaguine, S., 1995. Asymmetric Pd/Ag/Stainless steel catalytic membranes for methane steam reforming. Catal. Today 25, 327−332.

Spath P.L., Dayton, D.C., 2003. Preliminary Screening − Technical and Economic Assessment of Synthesis Gas to Fuels and Chemicals with Emphasis on the Potential for Biomass-Derived Syngas, National Renewable Energy Laboratory (NREL).

Udengaard, N.R., Hansen, J.H.B., Hanson, D.C., Stal, J.A., 1992. Sulfur passivated reforming process lowers syngas H2/CO ratio. Oil Gas J. 90, 62−67.

CHAPTER 6

The Future of Gasification

1 INTRODUCTION

The projections for the continued use of fossil fuels indicate that there will be at least another five decades of fossil fuel use (especially coal and petroleum) before biomass and other forms of alternative energy take a firm hold, although significant inroads are being made into the gasification of various feedstocks (Speight, 2008; Kumar et al., 2009; Pytlar, 2010; Speight, 2011a, b; Chhiti and Kemiha, 2013; Speight, 2013a, b). Furthermore, estimations that the era of fossil fuels (petroleum, coal, and natural gas) will be almost over when the cumulative production of the fossil resources reaches 85% of their initial total reserves (Hubbert, 1962) may or may not have some merit. In fact, the relative scarcity (compared to a few decades ago) of petroleum was real but it seems that the remaining reserves make it likely that there will be an adequate supply of energy for several decades (Martin, 1985; MacDonald, 1990; Banks 1992; Krey et al., 2009; Speight, 2011c, 2013a, 2013b, 2014). The environmental issues are very real and require serious and continuous attention.

Thus, gasification can be proposed as a viable alternative solution for energy recovery from a variety of feedstocks. On the other hand, the process still faces some technical and economic problems, mainly related to the highly heterogeneous nature of unconventional feedstocks such as biomass and municipal solid wastes and the relatively limited number of gasification plants worldwide based on this technology that have continuous operating experience under commercial conditions.

However, technologies which ameliorate the effects of fossil fuels combustion on acid rain deposition, urban air pollution, and global warming must be pursued vigorously (Vallero, 2008). There is a challenge that must not be ignored and the effects of acid rain in soil and water leave no doubt about the need to control its causes (Mohnen, 1988). Indeed, recognition of the need to address these issues is the driving force behind recent energy strategies as well as a variety of research and development programs (Stigliani and Shaw, 1990; United States Department of Energy, 1990; United States General Accounting Office, 1990).

While regulations on the greenhouse gas (GHG) carbon dioxide (CO_2) would be an immediate hurdle to deployment of coal plants, gasification plants are in the best position compared to other coal-based alternatives to capture carbon dioxide. However, with the continued uncertainty of carbon dioxide regulation, there is industry reluctance to make large investments in projects with high emissions of carbon dioxide since a cost-effective solution for reducing such emissions is not yet available. Nevertheless, the reduction in greenhouse gas emissions can be an enhancing factor for gasification in the long run because the carbon dioxide from a gasification plant is more amenable to capture.

As new technology is developed, emissions may be reduced by repowering in which aging equipment is replaced by more advanced and efficient substitutes. Such repowering might, for example, involve an exchange in which an aging unit is exchanged for a newer combustion chamber, such as the atmospheric fluidized bed combustor (AFBC) or the pressurized fluidized bed combustor (PFBC).

Indeed, recognition of the production of these atmospheric pollutants in considerable quantities every year has led to the institution of national emission standards for many pollutants. Using sulfur dioxide as the example, the various standards are not only very specific but will become more stringent with the passage of time. Atmospheric pollution is being taken very seriously and there is also the threat, or promise, of heavy fines and/or jail terms for any pollution-minded miscreants who seek to flaunt the laws (Vallero, 2008). Nevertheless, a trend to the increased use of fossil fuels will require more stringent approaches to environmental protection issues than we have ever known at any time in the past. The need to protect the environment is strong.

It is the purpose of this chapter to summarize gasification events and relate how they might be developed for future use.

2 ENVIRONMENTAL BENEFITS

The careless combustion of fossil fuels can account for the large majority of the sulfur oxides and nitrogen oxides released to the atmosphere.

Whichever technologies succeed in reducing the amounts of these gases in the atmosphere should also succeed in reducing the amounts of urban smog, those notorious brown and grey clouds that are easily recognizable at some considerable distances from urban areas, not only by their appearance but also by their odor.

$$SO_2 + H_2O \rightarrow H_2SO_4 \text{ (sulfurous acid)}$$
$$2SO_2 + O_2 \rightarrow 2SO_3$$
$$SO_3 + H_2O \rightarrow H_2SO_4 \text{ (sulfuric acid)}$$
$$2NO + H_2O \rightarrow HNO_2 + HNO_3 \text{ (nitrous acid + nitric acid)}$$
$$2NO + O_2 \rightarrow 2NO_2$$
$$NO_2 + H_2O \rightarrow HNO_3 \text{ (nitric acid)}$$

The most obvious issue with fossil fuel use relates to the effects on the environment. As technology evolves, the means to reduce the damage done by fossil fuel use also evolves and the world is on the doorstep of adapting to alternative energy sources. In the meantime, gasification offers alternatives to meet the demand for fuels of the future and to reduce the potentially harmful emissions.

Recent policy to tackle climate change and resource conservation, such as the Kyoto Protocol, the deliberations at Copenhagen in 2009, and the Landfill Directive of the European Union, stimulated the development of renewable energy and landfill diversion technology, so providing gasification technology development with a renewed impetus. However, even though they are the fastest growing source of energy, renewable sources of energy will still represent only 15% of the world energy requirements in 2035 (up from the current estimation of 10%) and divesting from fossil fuels does not mean an end to environmental emissions. Petroleum, tar sand bitumen, coal, natural gas, and perhaps oil shale will still be dominant energy sources − and will grow at a relatively robust rate over, at least, the next two decades. These estimates are a reality check on the challenge ahead for clean technologies if they are to make an impact in reducing greenhouse gas emissions and satisfy future energy demands (EIA, 2013).

Current awareness of these issues by a variety of levels of government has resulted, in the USA, of the institution of the Clean Fossil fuels Program to facilitate the development of pollution abatement technologies. And it has led to successful partnerships between government and industry (United States Department of Energy, 1993). In addition, there is the potential that new laws, such as the passage in 1990 of the Clean Air Act Amendments in the USA (United States Congress, 1990; Stensvaag, 1991), will be a positive factor and supportive of the controlled clean use of fossil fuels. However, there will be a cost but industry is supportive of the measure and confident that the goals can be met.

Besides fuel and product flexibility, gasification-based systems offer significant environmental advantages over competing technologies, particularly coal-to-electricity combustion systems. Gasification plants can readily capture carbon dioxide, the leading greenhouse gas, much more easily and efficiently than coal-fired power plants. In many instances, this carbon dioxide can be sold, creating additional value from the gasification process.

Carbon dioxide captured during the gasification process can be used to help recover oil from otherwise depleted oil fields. The Dakota Gasification plant in Beulah, North Dakota, captures its carbon dioxide while making substitute natural gas and sells it for enhanced oil recovery. Since 2000, this plant has captured and sent the carbon dioxide via pipeline to EnCana's Weyburn oil fields in Saskatchewan, Canada, where it is used for enhanced oil recovery. More than five million tons of carbon dioxide has been sequestered.

2.1 Carbon Dioxide

In a gasification system, carbon dioxide can be captured using commercially available technologies (such as the *water − gas shift reaction*) before it would otherwise be vented to the atmosphere. Converting the carbon monoxide to carbon dioxide and capturing it prior to combustion is more economical than removing carbon dioxide after combustion, effectively "de-carbonizing" or, at least, reducing the carbon in the synthesis gas.

Gasification plants manufacturing ammonia, hydrogen, fuels, or chemical products routinely capture carbon dioxide as part of the manufacturing process. According to the Environmental Protection Agency, the higher thermodynamic efficiency of the integrated gasification combined cycle (IGCC) minimizes carbon dioxide emissions relative to other technologies. IGCC plants offer a least-cost alternative for capturing carbon dioxide from a coal-based power plant. In addition, IGCC will experience a lower energy penalty than other technologies if carbon dioxide capture is required. While carbon dioxide capture and sequestration will increase the cost of all forms of power generation, an IGCC plant can capture and compress carbon dioxide at one-half the cost of a traditional pulverized coal plant. Other gasification-based options, including the production of motor fuels, chemicals, fertilizers or hydrogen, have even lower carbon dioxide capture and compression costs, which will provide a significant economic and environmental benefit in a carbon-constrained world.

2.2 Air Emissions

Gasification can achieve greater air emission reductions at lower cost than other coal-based power generation, such as supercritical pulverized coal. Coal-based IGCC offers the lowest emissions of sulfur dioxide nitrogen oxides and particulate matter (PM) of any coal-based power production technology. In fact, a coal IGCC plant is able to achieve low air-emissions rates that approach those of a natural gas combined cycle (NGCC) power plant. In addition, mercury emissions can be removed from an IGCC plant at one-tenth the cost of removal from a coal combustion plant. Technology exists today to remove more than 90% w/w of the volatile mercury from the synthesis gas in a coal-based gasification-based plant.

2.3 Solids Generation

During gasification, virtually all of the carbon in the feedstock is converted to synthesis gas. The mineral material in the feedstock separates from the gaseous products, and the ash and other inert materials melt and fall to the bottom of the gasifier as a non-leachable, glass-like solid or other marketable material. This material can be used for many construction and building applications. In addition, more than 99% w/w of the sulfur can be removed using commercially proven technologies and converted into marketable elemental sulfur or sulfuric acid.

2.4 Water Use

Gasification uses approximately 14 to 24% v/v less water to produce electric power from coal compared to other coal-based technologies, and water losses during operation are about 32 to 36% v/v less than other coal-based technologies. This is a major issue in many countries – including the USA – where water supplies have already reached critical levels in certain regions.

3 NOW AND THE FUTURE

Pyrolysis and gasification of fossil fuels, biomass materials, and wastes have been used for many years to convert organic solids and liquids into useful gaseous, liquid, and cleaner solid fuels. However, although gasification is old technology with respect to coal-based feedstock, it is a developing technology with respect to biomass and waste, since it has not yet been fully embraced on a larger commercial scale. Biomass feedstocks differ from conventional solid fuels such as coal-based fuels with regard to the volatile matter. The volatile matter in coal is less than 20% w/w, while it is up to 80% w/w in biomass materials (Quaak et al., 1999).

The renewed interest in biomass gasification is due to unpredictable fluctuations in the price of crude oil as well as unreliability in delivery of crude oil to the market (Speight, 2011c) and is evident from the number of commercial as well as developmental projects that are shaping up globally (Suresh Babu, 2006). While fossil fuels contaminate the atmosphere with sulfur dioxide, nitric oxide emissions using gasification instead of direct combustion can be reduced almost to zero and at the same time, the fuel gas quality can be improved to meet the requirements of different machinery, to produce heat and electricity. The gas produced from biomass gasification, or synthesis gas, is a mixture of carbon monoxide, hydrogen, and methane, together with carbon dioxide and nitrogen. The process also produces solid char and tars that would be liquid under normal ambient conditions. The proportions of these various constituents vary widely and will depend on the specific fuel source and operating conditions for conversion.

Gasification generally has a non-negative environmental image whereas combustion has a negative public image. The public view is not always based on objective analysis of emissions and other effects but it is still a very important factor. In fact, the technology of co-gasification can result in very clean power plants using a range of fuels but economic and environmental challenges may remain (Ricketts et al., 2002).

Gasification differs from more traditional energy-generating schemes in that it is not a combustion process, but rather a conversion process. Instead of the carbonaceous feedstock being wholly burned in air to create heat to raise steam which is used to drive turbines, the feedstock to be gasified is combined with steam and limited oxygen in a heated, pressurized vessel. The atmosphere inside the vessel is deficient in oxygen leading to a complex series of reactions of the feedstocks to produce synthesis gas. Moreover, using current technologies, the synthesis gas can be cleaned beyond environmental regulatory requirements (current and proposed), as demonstrated by currently commercial chemical production plants that require ultra-clean synthesis gas to protect the integrity of expensive catalysts. The clean synthesis gas can be combusted in turbines or engines using higher temperature (more efficient) cycles than the conventional steam cycles associated with burning carbonaceous fuels, allowing possible efficiency improvements. Synthesis gas can also be used in fuel cells and fuel cell-based cycles with yet even higher efficiencies and exceptionally low emissions of pollutants.

Furthermore, one of the major challenges of the 21st century is finding a way to meet national and global energy needs while minimizing the impact on the environment. There is extensive debate surrounding this issue, but there are certain areas of focus: (1) production of cleaner energy, both from conventional fuel sources and alternative technologies, (2) use of energy sources that are not only environmentally sound, but also economically viable, and (3) investment in a variety of technologies and resources to produce clean energy to meet energy needs. Gasification technologies will help to answer these challenges.

3.1 The Process

Gasification, a time-tested, reliable, and flexible technology, will be an increasingly important component of this new energy equation, even to the point of the evolution of the petroleum refinery as more gasification units are added to current refineries (Speight, 2011a). Any investment in gasification will yield valuable future returns in clean, abundant, and affordable energy from a variety of sources (Speight, 2008, 2011b).

Gasification is an environmentally sound way to transform any carbon-based material, such as coal, refinery byproducts, biomass, or even waste into energy by producing synthesis gas that can be converted into electricity and valuable products, such as transportation fuels, fertilizers, substitute natural gas, or chemicals (Chadeesingh, 2011; Speight, 2013a).

Gasification has been used on a commercial scale for approximately one hundred years by the coal, refining, chemical, and lighting industries. It is currently playing an important role in meeting energy needs in many countries and will continue to play an increasingly important role as one of the economically attractive manufacturing technologies that will allow production of clean, abundant energy. And, while gasification has typically been used in industrial applications, the technology is increasingly being adopted in smaller scale applications to convert biomass and waste-to-energy products.

Gasification is the cleanest, most flexible, and reliable way of using fossil fuels and a variety of other carbonaceous (carbon-containing) or hydrocarbonaceous (carbon-containing and hydrogen-containing) feedstocks. It can convert low-value materials into high-value products, such as chemicals and fertilizers, substitute natural gas, transportation fuels, electric power, steam, and hydrogen. The process can be used to convert biomass, municipal solid waste, and other materials (that are normally burned as fuel) into a clean gas. In addition, gasification provides the most cost-effective means of capturing carbon dioxide, a greenhouse gas, when generating power using fossil fuels as a feedstock. And most important for many countries dependent upon high-cost imported petroleum and natural gas from politically unstable regions of the world, gasification allows use of domestic resources to generate energy.

In fact, gasifiers can be designed to use a single material or a blend of feedstocks: (1) solids, such as coal, petroleum coke, biomass, wood waste, agricultural waste, household waste, and hazardous waste; (2) liquids, such as petroleum resids (including used/recovered road asphalts, tar sand bitumen and liquid wastes from chemical plants and other processing plants; (3) gas: natural gas or refinery/chemical off-gas.

The major sought-after products of gasification – synthesis gas and hydrogen – are dependent upon the specific gasification technology; smaller quantities of methane, carbon dioxide, hydrogen sulfide, and water vapor are also produced – typically, 70 to 85% of the carbon in the feedstock is converted into the synthesis gas. The ratio of carbon monoxide to hydrogen depends in part upon the hydrogen and carbon content of the feedstock and the type of gasifier used, but can also be adjusted downstream of the gasifier through use of catalysts. This inherent flexibility of the gasification process means that it can produce one or more products from the same process.

Another benefit of gasification is carbon dioxide removal in the synthesis gas clean-up stage using a number of proven commercial technologies (Mokhatab et al., 2006; Speight, 2007). In fact, carbon dioxide is routinely removed in gasification-based ammonia, hydrogen, and chemical manufacturing plants. Gasification-based ammonia plants already capture/separate approximately 90% v/v of the carbon dioxide and gasification-based methanol plants capture approximately 70% v/v of the carbon dioxide. The gasification process offers the most cost-effective and efficient means of capturing carbon dioxide during the energy production process.

Other byproducts include slag – a glass-like product – composed primarily of sand, rock, and minerals originally contained in the gasifier feedstock. This slag is non-hazardous and can be used in roadbed construction, cement manufacturing, or in roofing materials. Also, in most gasification plants, more than 99% w/w of the feedstock sulfur is removed and recovered either as elemental sulfur or sulfuric acid.

In addition, plasma gasification is increasingly being used to convert all types of waste, including municipal solid waste and hazardous waste, into electricity and other valuable products. Plasma is an ionized gas that is formed when an electrical charge passes through a gas. Plasma torches generate extremely high temperatures which, when used in a gasification plant, initiate and intensify the gasification reaction, increasing the rate of those reactions and making gasification more efficient. The plasma system allows different types of mixed feedstocks, such as municipal solid waste and hazardous waste, to avoid the expensive step of having to sort the feedstock by type before it is fed into the gasifier. These significant benefits make plasma gasification an attractive option for managing different types of wastes.

However, in contrast to coal, which is generally mined or imported thus large quantities are available from a single source or a number of closely located sources and supply has been reliable and predictable, neither biomass nor wastes are currently produced or naturally gathered at sites in quantities sufficient to fuel a modern large and efficient gasification plant, especially a plant that is designed to feed the product gases to a power generator. The disruption, transport, fuel use, and public opinion issues are not exactly in favor of gathering hundreds of tons of such fuels at a single location. Biomass or waste-fired power plants are therefore inherently limited in size and hence in efficiency, labor costs per unit electricity produced, and in other economies of scale. The production rates of municipal refuse follow reasonably predictable patterns over time periods of a few years. Recent experience with the very limited current *biomass for energy* harvesting has shown unpredictable variations in harvesting capability with long periods of zero production over large areas during wet weather and the foot and mouth outbreak.

On the one hand, combining biomass, refuse, and coal overcomes the potential unreliability of biomass, the potential longer-term changes in refuse, and the size limitation of a power plant using only waste and/or biomass. This also allows realization of the benefit for production and the price for electricity from biomass and the gate fee associated with waste. If the power plant is gasification based, rather than direct combustion based, further benefits may be available (Speight, 2008, 2013b).

Thus, there is a need to develop the infrastructure for the use of biomass and waste to generate the gaseous products necessary for further use. This will eventually happen but at current rates of development it may be a decade or so before the infrastructure is in place (Ciferno and Marano, 2002; Obernberger and Thek, 2008).

3.2 Refinery of the Future

As we enter the 21st century, petroleum refining technology is experiencing great innovation driven by the increasing supply of heavy oils with decreasing quality and the fast increases in the demand for clean and ultra-clean vehicle fuels and petrochemical raw materials (Speight and Ozum, 2002; Hsu and Robinson, 2006; Gary et al., 2007; Speight, 2014). As feedstocks for refineries change, there must be an accompanying change in refinery technology. This means a movement from conventional means of refining heavy feedstocks using (typically) coking technologies to more innovative processes (including hydrogen management) that will produce the ultimate amounts of liquid fuels from the feedstock and maintain emissions within environmental compliance (Penning, 2001; Lerner, 2002; Davis and Patel, 2004; Speight, 2008, 2011a).

Over the next 20 to 30 years, the evolution of petroleum refining and the current refinery configuration (Speight, 2014) will focus primarily on process modification with some new innovations coming on-stream. The industry will move predictably toward: (1) deep conversion of heavy feedstocks, (2) higher hydrocracking and hydrotreating capacity, and (3) more efficient processes.

High conversion refineries will move to gasification of feedstocks for the development of alternative fuels and to enhance equipment usage. A major trend in the refining industry market demand for refined products will be in synthesizing fuels from simple basic reactants (e.g., synthesis gas) when it becomes uneconomical to produce super clean transportation fuels through conventional refining processes. Fischer − Tropsch plants together with IGCC systems will be integrated with or even into refineries, which will offer the advantage of high quality products (Stanislaus et al., 2000).

A gasification refinery would have, as the center piece, gasification technology as is the case with the Sasol refinery in South Africa (Couvaras, 1997). The refinery would produce synthesis gas (from the carbonaceous feedstock) from which liquid fuels would be manufactured using Fischer − Tropsch synthesis technology.

In fact, gasification to produce synthesis gas can proceed from any carbonaceous material, including biomass. Inorganic components of the feedstock, such as metals and minerals, are trapped in an inert and environmentally safe form as char, which may have use as a fertilizer. Biomass gasification is therefore one of the most technically and economically convincing energy possibilities for a potentially carbon neutral economy.

A modified version of steam reforming known as autothermal reforming, which is a combination of partial oxidation near the reactor inlet with conventional steam reforming further along the reactor, improves the overall reactor efficiency and increases the flexibility of the process. Partial oxidation processes using oxygen instead of steam also found wide application for synthesis gas manufacture, with the special feature that they could utilize low-value feedstocks such as heavy petroleum residua. In recent years, catalytic partial oxidation employing very short reaction times (milliseconds) at high temperatures (850 to 1000°C, 1560 to 1830°F) is providing still another approach to synthesis gas manufacture (Hickman and Schmidt, 1993).

As petroleum supplies decrease, the desirability of producing gas from other carbonaceous feedstocks will increase, especially in those areas where natural gas is in short supply. It is also anticipated that the costs of natural gas will increase, allowing coal gasification to compete as an economically viable process. Research in progress on a laboratory and pilot-plant scale should lead to the invention of new process technology by the end of the century, thus accelerating the industrial use of coal gasification.

The conversion of the gaseous products of gasification processes to synthesis gas, a mixture of hydrogen (H_2) and carbon monoxide (CO), in a ratio appropriate to the application, needs additional steps, after purification. The product gases – carbon monoxide, carbon dioxide, hydrogen, methane, and nitrogen – can be used as fuels or as raw materials for manufacture of chemicals or fertilizers.

3.3 Economic Aspects

While a gasification plant is capital intensive (like any manufacturing unit), its operating costs can be lower than many other manufacturing processes or coal combustion plants. Because a gasification plant can use low-cost feedstocks, such as petroleum coke or high-sulfur coal, converting them into high-value products, it increases the use of available energy in the feedstocks while reducing disposal costs. Ongoing research, development, and demonstration of investment efforts show potential to substantially decrease current gasification costs even further, driving the economic attractiveness of gasification.

In addition, gasification has a number of other significant economic benefits: (1) the principal gasification byproducts – sulfur, sulfuric acid, and slag – are marketable, (2) gasification can produce a number of high-value products at the same time (*co-generation* or *polygeneration*), helping a facility offset its capital and operating costs and diversify its risks, (3) gasification offers wide feedstock flexibility – a gasification plant can be designed to vary the mix of the solid feedstocks or run on natural gas or liquid feedstocks when desirable, and (4) gasification units require less emission control equipment because they generate fewer emissions, further reducing the plant's operating costs.

Investment in gasification injects capital into the economy (by building large-scale plants using domestic labor and suppliers) and creates domestic jobs (construction to build, well-paid jobs to run) that cannot be outsourced to overseas workers.

4 MARKET DEVELOPMENTS

The future depends very much on the effect of coal gasification processes on the surrounding environment. It is these environmental effects and issues that will direct the success of gasification.

Clean coal technologies are a new generation of advanced coal utilization processes that are designed to enhance both the efficiency and the environmental acceptability of coal extraction, preparation, and use. These technologies reduce emissions, reduce waste, and increase the amount of energy gained from coal. The goal of the program was to foster development of the most promising clean coal technologies such as improved methods of cleaning coal, fluidized bed combustion, integrated gasification combined cycle, furnace sorbent injection, and advanced flue-gas desulfurization.

In fact, there is the distinct possibility that within the foreseeable future the gasification process will increase in popularity in petroleum refineries – some refineries may even be known as gasification refineries (Speight, 2011b).

The manufacture of gas mixtures of carbon monoxide and hydrogen has been an important part of chemical technology for about a century. Originally, such mixtures were obtained by the reaction of steam with incandescent coke and were known as *water gas*. Eventually, steam reforming processes, in which steam is reacted with natural gas (methane) or petroleum naphtha over a nickel catalyst, found wide application for the production of synthesis gas.

In a gasifier, the carbonaceous material undergoes several different processes: (1) pyrolysis of carbonaceous fuels, (2) combustion, and (3) gasification of the remaining char. The process is very dependent on the properties of the carbonaceous material and determines the structure and composition of the char, which will then undergo gasification reactions.

The forecast for growth of gasification capacity focuses on two areas: large-scale industrial and power generation plants and the smaller scale biomass and waste-to-energy area.

Worldwide industrial and power generation gasification capacity is projected to grow by 70% by 2015, with 81% of the growth occurring in developing markets. The prime movers behind this expected growth are the chemical, fertilizer, and coal-to-liquids industries in China, tar sands in Canada, polygeneration (hydrogen and power or chemicals) in the USA, and refining in Europe. In fact, China has focused on gasification as part of the overall energy strategy. The industrial and power gasification industry in the USA faces a number of challenges, including rising construction costs and uncertainty about policy incentives and regulations. Despite these challenges, the industrial and power gasification capacity in the USA is expected to grow. A number of factors will contribute to this expansion: (1) volatile oil and natural gas prices will make low-cost and abundant domestic resources with stable prices increasingly attractive as feedstocks, and (2) gasification processes will be able to comply with more stringent environmental regulations because their emission profiles are already substantially less than more conventional technologies.

In fact, there is a growing consensus that carbon dioxide management will be required in power generation and energy production. Since the gasification process allows carbon dioxide to be captured in a cost-effective and efficient manner, it will be an increasingly attractive choice for the continued use of fossil fuels. The greatest area of growth in terms of the number of plants in the USA is likely to be in the biomass and waste-to-energy gasification areas. Because they are smaller in scale, these plants are easier to finance, easier to permit, and take less time to construct. In addition, municipal and state restrictions on landfills and incineration and a growing recognition that these materials contain valuable sources of energy are driving the demand for these plants.

Furthermore, a number of factors will contribute to the growing interest in waste and biomass gasification: (1) restrictions on landfill space, (2) efforts to reduce costs associated with waste management, (3) growing recognition that waste and biomass contain unused energy that can be captured and converted into energy and valuable products, and (4) ability to use non-food biomass materials and convert them into valuable energy products

5 OUTLOOK

The ever-increasing global energy demand and the fast depleting fossil fuels have shifted focus on sustainable energies such as biomass and waste in the recent past. The importance of the gasification of such alternative feedstocks cannot be under-appreciated as potential sources of sustainable energy to meet the energy demand of future generations. The various technologies that are in practice at the commercial and pilot scale, with respect to bubbling, circulating fluidized beds and dual fluidized beds, are being developed for feedstocks other than coal (Bridgwater, 2002; Speight, 2008, 2011a,b,c, 2014).

In fact, gasification could now be proposed as a viable alternative solution for biomass waste treatment with energy recovery. However, the gasification process still faces some technical and economic problems, mainly related to the highly heterogeneous nature of feeds like municipal solid wastes and the relatively limited number of plants (approximately 100) worldwide based on this technology that have continuous operating experience under commercial conditions. However, the issues associated with biomass are not insurmountable. The high moisture content and the high mineral matter content can cause ignition and combustion problems but high moisture content biomass is more appropriate for a wet conversion process, such as fermentation, i.e., processes involving biochemically mediated reactions, whereas low moisture content biomass is more economically suited to a conversion process, such as gasification (Oliveira and Franca, 2009).

In the aggressive working environment of municipal solid waste management, with its uncompromising demand for reasonable cost, high reliability, and operational flexibility, it could be premature to identify gasification as the thermal processing strategy of the future or even as a strong competitor for combustion systems, at least for all sizes of waste-to-energy plants.

The success of any advanced thermal technology is determined by its technical reliability, environmental sustainability, and economic convenience. The technical reliability and the on-line availability is supported by years of successful continuous operations of about one hundred gasification-based <u>waste to energy</u> plants, mainly in Japan but now also in Korea and Europe. The environmental performance is one of the greatest strengths of gasification technology, which often is considered a sound response to the increasingly restrictive regulations applied around the world: independently verified emissions tests indicate that gasification is able to meet existing emissions limits and can have a great effect by reducing landfill disposal.

Economic aspects are probably the crucial factor for a relevant market penetration, since gasification-based WtE tends to have ranges of operating and capital costs higher than those of conventional combustion-based waste-to-energy (in the order of about 10%), mainly as a consequence of the ash melting system or, in general, of the added complexity of the technology.

The evidence of the last decade or so indicates an advantage for gasification plants processing less than about 120 000 tons per year. The greatest technical challenges to overcome for a wider market penetration of commercial advanced gasification technologies still appears to be that of an improved and cheaper synthesis gas cleaning, able to conveniently meet defined specifications and to obtain higher electric energy conversion efficiencies. It is essential that the performance and experience from several commercial waste gasifiers in operation will point to the gasification process as a strong competitor of conventional moving grate or fluidized bed combustion systems.

REFERENCES

Babu, S.P., 2006. Thermal gasification of biomass. Proceedings. Workshop No. 1: Perspectives on Biomass Gasification. IEA Bioenergy Agreement, Task 33. International Energy Agency, Paris, France.

Banks, F.E., 1992. Some aspects of natural gas and economic development − a short note. OPEC Bull. 16 (2), 235−240.

Bridgwater, A.V., 2002. The Future for Biomass Pyrolysis and Gasification: Status, Opportunities and Policies for Europe. Contract No: 4.1030/S/01-009/2001. Energy Framework Program, European Union, Directorate-General for Transport and Energy, Brussels, Belgium.

Chadeesingh, R., 2011. The fischer − tropsch process. In: Speight, J.G. (Ed.), The Biofuels Handbook. The Royal Society of Chemistry, London, UK, pp. 476−517. (Part 3, Chapter 5).

Chhiti, Y., Kemiha, M., 2013. Thermal conversion of biomass, pyrolysis and gasification: a review. Int. J. Eng. Sci. 2 (3), 75−85.

Ciferno, J.P., Marano, J.J., 2002. Benchmarking Biomass Gasification Technologies for Fuels, Chemicals and Hydrogen Production. National Energy Technology Laboratory, U.S. Department of Energy, Washington, DC.

Couvaras, G., 1997. Sasol's slurry phase distillate process and future applications. Proceedings Monetizing Stranded Gas Reserves Conference, Houston, December 1997.

Davis, R.A., Patel, N.M., 2004. Refinery hydrogen management. Petroleum Technol. Q. Spring, 29–35.

EIA, 2013. International Energy Outlook 2013: World Energy Demand and Economic Outlook. International Energy Agency, Paris, France. <http://www.eia.gov/forecasts/ieo/world.cfm> (accessed 13.09.13).

Gary, J.H., Handwerk, G.E., Kaiser, M.J., 2007. Petroleum Refining: Technology and Economics 5th Edition. CRC Press, Taylor & Francis Group, Boca Raton, FL.

Hickman, D.A., Schmidt, L.D., 1993. Production of syngas by direct catalytic oxidation of methane. Science 259, 343.

Hsu, C.S., Robinson, P.R., 2006. Practical Advances in Petroleum Processing, Volumes 1 and 2. Springer, New York.

Hubbert, M.K., 1962. Energy Resources. National Academy of Sciences, Washington, DC, Report to the Committee on Natural Resources.

Krey, V., Canadell, J.G., Nakicenovic, N., Abe, Y., Andruleit, H., Archer, D., et al., 2009. Gas hydrates: entrance to a methane age or climate threat? Environ. Res. Lett. 4 (3), 034007.

Kumar, A., Jones, D.D., Hanna, M.A., 2009. Thermochemical biomass gasification: a review of the current status of the technology. Energies 2, 556–581.

Lerner, B., 2002. The future of refining. Hydrocarbon Eng. September.

MacDonald, G.J., 1990. The future of methane as an energy resource. Annu. Rev. Energy 15, 53–83.

Martin, A.J., 1985. The prediction of strategic reserves. In: Niblock, T., Lawless, R. (Eds.), Prospects for the World Oil Industry. Croom Helm Publishers, Beckenham, Kent (Chapter 1).

Mohnen, V.A., 1988. The challenge of acid rain. Sci. Am. 259 (2), 30–38.

Mokhatab, S., Poe, W.A., Speight, J.G., 2006. Handbook of Natural Gas Transmission and Processing. Elsevier, Amsterdam, Netherlands.

Obernberger, I., Thek, K., 2008. Combustion and gasification of solid biomass for heat and power production in europe – state-of-the-art and relevant future developments. Proceedings Eighth European Conference on Industrial Furnaces and Boilers. Vilamoura, Portugal. March 25–28. INFUB, Rio Tinto, Portugal.

Oliveira, L.S., Franca, A.S., 2009. From solid bio-waste to liquid biofuels. In: Ashworth, G.S., Azevedo, P. (Eds.), Agricultural Wastes. Nova Science Publishers Inc., Hauppauge, New York.

Penning, R.T., 2001. Petroleum refining: a look at the future. Hydrocarbon Process. 80 (2), 45–46.

Pytlar, T.S. Jr., 2010. Status of Existing Biomass Gasification and Pyrolysis Facilities in North America. Paper No. NAWTEC 18-3521. Proceedings. 18th Annual North American Waste-to-Energy Conference, NAWTEC18, Orlando, Florida. May 11–13.

Quaak, P., Knoef, H., Stassen, H., 1999. Energy from Biomass: A Review of Combustion and Gasification Technologies. Technical Paper No. 422. World Bank, Washington, DC.

Ricketts, B., Hotchkiss, R., Livingston, W., Hall, M., 2002. Technology status review of waste/biomass co-gasification with coal. Proceedings Inst. Chem. Eng. Fifth European Gasification Conference. Noordwijk, Netherlands. April 8–10.

Speight, J.G., Ozum, B., 2002. Petroleum Refining Processes. Marcel Dekker Inc., New York.

Speight, J.G., 2007. Natural Gas: A Basic Handbook. GPC Books, Gulf Publishing Company, Houston, Texas.

Speight, J.G., 2008. Synthetic Fuels Handbook: Properties, Processes, and Performance. McGraw-Hill, New York.

Speight, J.G., 2011a. The Refinery of the Future. Gulf Professional Publishing, Elsevier, Oxford, UK.

Speight, J.G., 2011b. The Biofuels Handbook. Royal Society of Chemistry, London, UK.

Speight, J.G., 2011c. An Introduction to Petroleum Technology, Economics, and Politics. Scrivener Publishing, Salem, MA.

Speight, J.G., 2013a. The Chemistry and Technology of Coal 3rd Edition. CRC Press, Taylor & Francis Group, Boca Raton, FL.

Speight, J.G., 2013b. Coal-Fired Power Generation Handbook. Scrivener Publishing, Salem, MA.

Speight, J.G., 2014. The Chemistry and Technology of Petroleum, 5th Edition. CRC Press, Taylor & Francis Group, Boca Raton, FL.

Stanislaus, A., Qabazard, H., Absi-Halabi, M., 2000. Refinery of the Future. Proceedings. 16th World Petroleum Congress, Calgary, Alberta, Canada. June 11 − 15.

Stensvaag, J.-M., 1991. Clean Air Act Amendments: Law and Practice. John Wiley and Sons Inc, New York.

Stigliani, W.M., Shaw, R.W., 1990. Energy use and acid deposition: the view from europe. Annu. Rev. Energy 15, 201−216.

United States Congress., 1990. Public Law 101-549. An Act to Amend the Clean Air Act to Provide for Attainment and Maintenance of Health Protective National Ambient Air Quality Standards, and for Other Purposes. November 15.

United States Department of Energy. 1990. Gas Research Program Implementation Plan. DOE/FE-0187P. United States Department of Energy, Washington, DC. April.

United States Department of Energy., 1993. Clean Fossil Fuels Technology Demonstration Program. DOE/FE-0272. United States Department of Energy, Washington, DC. February.

United States General Accounting Office., 1990. Energy Policy: Developing Strategies for Energy Policies in the 1990s. Report to Congressional Committees. GAO/RCED-90-85. United States General Accounting Office, Washington, DC. June.

Vallero, D., 2008. Fundamentals of Air Pollution 4th Edition. Elsevier, London, UK.

Note: This content is available in its entirety online at
store.elsevier.com/product.jsp?isbn=9780127999111 (click the
Resources tab at the bottom of the page).

Gasification of Unconventional Feedstocks. DOI: http://dx.doi.org/10.1016/B978-0-12-799911-1.00017-0